No from

caro

Hugh

do

From

londres

7. 12. 7

Man Flies

THE STORY OF ALBERTO SANTOS-DUMONT

MASTER OF THE

BALLOON

Man Flies

THE STORY OF ALBERTO SANTOS-DUMONT

MASTER OF THE

BALLOON

CONQUEROR OF THE AIR

" I am serving my apprenticeship to the
métier de l'oiseau"

NANCY WINTERS

BLOOMSBURY

For "Little Santos"

con saudades

and

for Condor & Condor

who taught me to fly

First published in Great Britain 1997
Bloomsbury Publishing Plc, 38 Soho Square,
London W1V 5DF

Picture sources

Aeronautica: pages 89 *top,* 121; Pierre Cardin: page 41 *top*; Cartier: pages 41, 42, 73, 117, 118, 119, 126; *Copacabana Picture Book*: pages 141,142; *Dans l'air*: pages 14, 27, 29 *top*, 50, 58, 63, 64, 67, 68, 83, 84, 85; Dumont-Villares, *Santos Dumont 1898-1910*: pages 13 *top & bottom*, 29 *bottom*, 31, 89 *bottom*, 98; Hulton Getty: pages 52-3, 71, 143; *Illustrated London News*: pages 24, 25; F.H. La Costa: pages 54, 88, 133, 138; *A Passion for Wings*, pages 40, 56-7, 80, 97, 107 *bottom*; *Revue Ampère*: pages 44, 61, 90, 95, 96, 122 *top*, 136-7, 139; Royal Aeronautical Society: pages 37, 47, 60; Judy Rudoe, *Cartier 1900-1939*: pages 113, 114; Science and Society Picture Library: pages 17 *inset*, 23, 32, 34, 38,72, 74,78 *bottom*, 103, 104, 106, 123 *top*, 129, 130, 131; *A Study in Obsession*: pages 20, 26, 45, 75, 77, 87, 91, 99, 101, 105 *bottom*, 135; TRH: pages 51, 59, 91, 107 *top*, 108, 123, 125, 127; Liebig/Bridgeman Picture Library: page 34 *inset*; Roger Viollet: pages 28, 48-9, 92, 111, 124.

A CIP catalogue record for this book is available from the British Library

ISBN 0 7475 3596 5

10 9 8 7 6 5 4 3 2 1
Studio photography by Andrea Heselton
Designed by Bradbury and Williams
Printed by Artegrafica S.p.A., Verona, Italy

CONTENTS

Author's Note

Thisis not a history of flight. Nor even of ballooning. It is the story of one small, courageous, stubborn, stylish and, ultimately, tragic man. It is not so much a story of science as a story of dreams. For me, it began with a Cartier watch. It was the height of the extravagant eighties and I was being paid to be a high-flying travel journalist. The quartz watch to which I had treated myself ("Guaranteed to run for a lifetime") had stopped dead after just a few hours.

"Whose lifetime is this thing guaranteed for?" I asked Bill San Filippo, the salesman at Cartier New York, from whom I had already bought a fountain pen.

"Well," he soothed, "there must be something wrong with this one. Have another." The customer is generally right at Cartier, and I think he was also grooming me for higher things.

But the replacement also, alas, died after a few hours. (It wasn't then generally known that some people have this effect on quartz watches.)

"I can't understand it," said the by now embarrassed Bill. He offered to select yet another replacement watch and to monitor it personally for twenty-four hours.

Just back from gliding in Iceland (a terrifying experience), I was on the point of departing on another of the kind of trips that I believe had convinced him (incorrectly) that I was a wealthy eccentric. "Don't I get a loaner, then?" I teased. "As when one's sports car needs repair?"

There was just an instant's hesitation. Then he handed over — without asking for a receipt of any kind — what became, on first sight, one of the delights of my life A Santos Sport. I loved the curve of it. The chunkiness. The layering of the case. The little screws and the mix of steel and gold. I felt dashing wearing it without quite knowing why. (See page 119.) And I have rarely taken it off since.

Over the next six or seven years, a series of coincidences gradually turned what had begun as a bit of a lark into something more serious. My watch, I knew, had been named after an aviator, Alberto Santos-Dumont, but I had not paid the topic much further attention. Assignments took me, several times, to Brazil, where (still larkily) I made a point of having my watch photographed next to the statue of its inspiration at Rio's Santos-Dumont Airport. In São Paulo I was taken to the Museu da Aeronautica at Ibirepuera where, a bit more soberly, I viewed the Santos-Dumont display and marvelled at the tiny basket in which he had ascended in his first balloon, the Brazil; a copy of and receipt for the original watch; and a life-size model of the extraordinary Heath-Robinson-like contraption in which he made what all the world for several years believed was the first validated flight. I walked along the long, broad beach at Guarujá and was moved by its beauty – despite the hotels that have grown up along it – before I ever learned that this was where he spent his declining days and final hours.

As the excesses of the eighties waned, I eschewed my own high flying in favour of the more grounded life of a novelist in London. But Alberto Santos-Dumont followed me. His piercing eyes stared out from the pages of his book *Dans l'air*, which I came upon in a second-hand bookshop. The very veranda on which as a child he read Jules Verne, and built his first little hot-air balloons, appeared in a portfolio of photographs of the coffee plantation where he grew up, found in a catalogue of old cookery books. Some original sketches by his friend SEM and a tiny charm – a replica of the airship with which he won the Deutsch Prize – turned up in an auction at Bonham's. (And I, eccentric possibly but still not wealthy, nervously bid for – and now own – them both.)

I began to formalize my research. I learned the basic story of dreams, daring, persistence and finally victory, too soon and cruelly snatched away. Of a man forgotten because of his generosity, and resurrected because of his style.

Digging deeper, I rediscovered photographs and facts long hidden in museums and dusty archives, and spoke with balloonists, collectors and what "Little Santos" would have referred to as interested parties. Throughout, I felt his vision inspiring me as it had once inspired not only his colleagues but the world.

In the course of my exploring his importance as a link in the history of aviation, there came two moments I shall always remember.

The first was when I heard that Charles Lindbergh, who had requested that Santos-Dumont (who wept when declining, on account of his illness) attend the celebration dinner held for him by the Aero Club, years later himself dined, on the eve of their historic 1968 first flight around the moon, with the astronauts of Apollo 8.

The second occurred as I sat surrounded, indeed almost hemmed in, by the stacks of files at the Royal Aeronautical Society's Aviation Library. Suddenly, something made me lift my head and, looking up, I saw a bird fly, easily, lightly, jauntily, past the window and soar out over Hyde Park.

Nancy Winters *London, August 1997*

Foreword

I t is 1973 and the snubby nose of a small silvery airship emerges from the shadows of the vast airship hangars at Cardington in Bedfordshire, England. Built by a band of dedicated enthusiasts in the spirit of adventure, and just big enough to carry two adults aloft, it bears on its bulbous flanks the name of its inspiration and their hero – Santos-Dumont.

Born one hundred years earlier, Alberto Santos-Dumont was to the Belle Époque what John Glenn and Neil Armstrong are to our generation. Perhaps not so obviously of the "right stuff", this dapper "Little Icarus" from Brazil achieved international renown through his pioneering exploits high over the streets of Paris. Driven by an absolute and unquestioning conviction that man could and would fly, he took the wayward balloon and tamed it to create a series of aerial runabouts – little "dirigibles" or airships – with names such as the Racer and the Stroller.

His successes, most notably in winning the 1901 Deutsch Prize steering a circuit around the Eiffel Tower, brought him prizes and celebrity even greater than Lindbergh would know.

Yet Santos-Dumont's achievements have been obscured by the twentieth century's headlong rush for progress. His greatest claim to fame, as the first man to fly a powered aircraft, was eclipsed by the Wright brothers, of whose work the world still knew nothing when he was making his first tentative hops.

Nowadays the name Santos-Dumont means little to people outside the narrow confines of the airship world, except perhaps to those who know of the famous Cartier watches he inspired. Indeed, when the little 1970s airship turned up in Los Angeles, the American sponsor commented, "Why choose the name of some dead guy?" It is in Brazil that his status is most vehemently upheld (and the airport in Rio is named after him).

Nancy Winters's excellent new appraisal of his life and achievements does much to redress the balance. It is a timely reminder too, for as we approach the millennium, balloonists are attempting to fly non-stop around the world and the airship is flourishing, with more flying now than at any time in the last fifty years. And these are not the great metal-framed leviathans envisaged by Count von Zeppelin. No, these are fleets of smaller craft fulfilling genuine commercial roles – little ships of the sky which my hero, Alberto Santos-Dumont, inspired and which he would have no trouble in recognizing.

John Christopher
Balloonist and Editor of Aerostat *magazine*

CHAPTER ONE

Navigating with Hector Servadoc

"I, TOO, DESIRED TO GO BALLOONING."

Although now known mainly for the watch created for him by Louis Cartier, the Brazilian balloonist Alberto Santos-Dumont once enjoyed a fame even greater than that of Lindbergh or the early astronauts. Internationally acclaimed as the first man to fly, he was fêted for several years in Europe and South America, as well as in the United States (where he was received at the White House by Teddy Roosevelt), before learning that the Wright brothers, whose early efforts had been discounted, had actually preceded him.

The story of how this brilliant, colourful and eccentric pioneer slipped through the cracks of history — and resurfaced only because of the watch itself — is glamorous, tragic and inspiring.

The dreamy youngest child of a wealthy plantation owner known as the Coffee King of Brazil, Alberto Santos-Dumont spent his childhood in a fantasy of flight and machines.

Both his parents were first-generation Brazilians. His mother, Francisca dos Santos, a devout and undemonstrative woman, was descended from Brazilian aristocracy. Her family, escorted by British frigates, had arrived from Portugal with its king, Dom João, in flight from Napoleon.

High white collars were also favoured by Dona Francisca.

His father's forebears had been French jewellers, emigrating a few years later; his father, Henrique Dumont, an engineer and adventurer of sorts, who was rumoured to have climbed Mont Blanc alone in his youth, was educated in Paris. After a number of unsuccessful business ventures, mostly backed or instigated by his wealthy father-in-law, he managed to acquire a vast tract of land which, calling on the skills from his earlier occupations – civil engineering, steamboating, railroad building – he developed into the largest coffee plantation in the country.

Alberto's father was rumoured to have climbed Mont Blanc.

Young Alberto, born on his father's forty-first birthday in 1873, became the family pet. Indulged by his much older brothers (two) and sisters (four), and left to his own devices, he devoured the novels of Jules Verne (which no

At seven, Alberto was driving locomotives on the 60-mile (private) railway.

one told him weren't true).

"With Captain Nemo and his shipwrecked guests I explored the depths of the sea in that first of all submarines, the *Nautilus*. With Phileas Fogg I went round the world in eighty days, in Screw Island and the Steam House my boyish faith leaped out to welcome the automobiles which, in those days, had not yet a name," he wrote years later.

But from the very first, it was ballooning that had his heart.

"With Hector Servadoc," he confessed, "I navigated the air. I, too, desired to go ballooning. In the long, sun-bathed Brazilian afternoons, when the hum of insects, punctuated by the far-off cry of some bird, lulled me, I would lie in the shade of the veranda and gaze into the fair sky of Brazil, where the birds fly so high and soar with such ease on their great out-stretched wings, where the clouds mount so gaily in the pure light of day, and you have only to raise your eyes to fall in love with space and freedom. So, musing on the exploration of the aerial ocean, I, too, devised airships and flying machines in my imagination."

He was not only imagining, though, those long hours on the veranda watching the condors and dreaming of "sailing" the sky. He made little fire-balloons of silk paper — which he called Montgolfiers after the eighteenth-century balloon pioneers — and sent them up in fleets against the bonfire-lit sky on St John's Eve, and light planes from straw, powered by springs of twisted rubber, "waiting for something better".

But he knew enough to keep his dreams to himself, for to acknowledge them would have been "to stamp oneself as unbalanced and visionary".

Instead, while his father and brothers made inspection trips on horseback, he haunted the processing works, where he learned both to operate and to repair the equipment. At seven, he was driving traction engines; at twelve, the Baldwin loco-motives of the plantation's 60-mile railway.

What might have seemed like mere boyish adventures, how-ever, were actually important learning experiences that great-ly influenced his future inventions. Playing with the coffee engines, as he described it, he became familiar not only with the workings of the various machines — pulpers, separators, skinners and huge ventilators — but also with their strengths and weaknesses. It was this that turned him against all agitating

devices, and led to his later use of rotary-action powering.

He was educated at home in his early years by his favourite sister, Virginia. His main companions were the children of the staff and, although as an outsider he was taunted for it, he refused to hide his dreams and goals from them.

"I remember how my comrades used to tease me at our

In the sun-bathed afternoons he would lie on the veranda and devise "Montgolfiers".

game of 'Pigeon Flies!'" he reminisced years later in his charming book *Dans l'air*. "All the children gather round a table and the leader calls out 'Pigeon flies! Hen flies! Crow flies! Bee

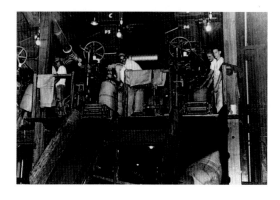

Playing with the coffee engines greatly influenced his future.

flies!' and so on; and at each call we were supposed to raise our fingers. Sometimes, however, he would call out: 'Dog flies! Fox flies!' or some other like impossibility, to catch us. If anyone raised a finger, he was made to pay a forfeit. Now my playmates never failed to wink and smile mockingly at me when one of them called 'Man flies!' for at the word I would always lift my finger very high, as a sign of absolute conviction; and I refused with energy to pay the forfeit. The more they laughed at me, the happier I was, hoping that someday the laugh would be on my side."

Thus he grew up. Protected, enchanted, determined and, basically, misled.

When he was eighteen, it all suddenly ended.

His father, thrown from his horse and partially paralysed, went to Paris for medical treatment, taking Alberto along. When he realized he wouldn't recover, he sold the plantation, divided up the shares and, not long afterwards, died.

Within the year, the young man found himself on his own in Paris with a vast fortune at the height of the Belle Époque.

CHAPTER TWO

Steerable Balloons and Automobiles

"FOLLOW MY DIRECTIONS AND DO NOT
CONCERN YOURSELVES WITH THEIR
PRACTICABILITY."

Paris at the turn of the century was an exciting place to be. Innovation was in the air. The Eiffel Tower had just been built, radioactivity and X-rays just discovered, and the wireless just invented. The first few automobiles were jumping down the avenues and Paris's Grande Exposition was being planned.

Picasso, Cézanne, Matisse and Monet were making their names, as were Sarah Bernhardt, Nellie Melba and Colette (while Oscar Wilde, under the name of Sebastian Melmoth, was trying to hide his).

Proust was in his cork-lined room.

Toulouse Lautrec was drinking absinthe at the Moulin Rouge.

The great courtesans, exquisitely dressed, could be seen in the restaurants and cafés at night and, each morning, walking

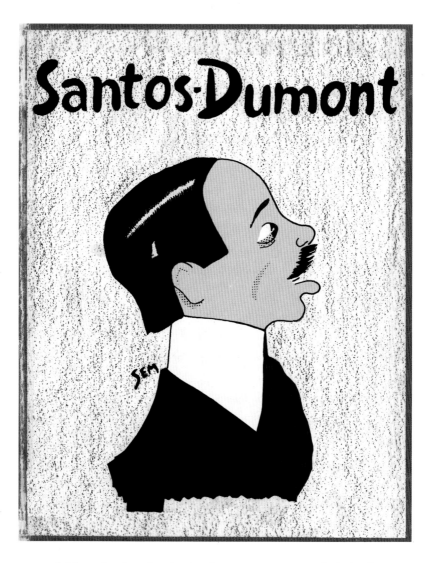

Although a bit of a dandy, he was not just another playboy.

their little dogs in the Bois de Boulogne (where the infant Diana Vreeland, who had an excellent view from her pram, was so impressed by the beauty of their shoes that she decided to devote her life to fashion).

But the wealthy young Brazilian wasn't interested in all this. The "new things" he reported himself to be going to Paris to see were "steerable balloons and automobiles".

Dapper, a bit of a dandy, with slick, centreparted hair and a splendid moustache, the slight but daring new arrival seemed at first like just another playboy, frequenting Maxim's and launching the dangerous sport of motor-tricycle racing in the Parc des Princes. But, despite appearances, he was no dilettante. As the Parisians were soon to learn, he had something much more important on his mind. He was going to conquer the air.

He set about making his dream come true. And got the first big shock of his life.

To his amazement he found that, except in the stories of Jules Verne, ballooning had made little real progress since the experiments of the Montgolfier brothers over 100 years before. Despite the experiments of Giffard and Lilienthal and others using everything from paddle wheels to airborne horses, no one had yet devised a successful dirigible, or steerable, balloon.

Free flight, at the mercy of the wind, was still the order of the day and no one took it much more seriously than had Dr

Johnson who, in 1784, commented, "These vehicles can serve no use until we can guide them. I had rather now find a medicine that can cure an asthma."

The activities of the aeronauts, as they were then called, were confined mainly to daredevil stunts at fairs, ill-fated expeditions to exotic places, and keeping what little knowledge they had to themselves. So-called "captive balloons" (tethered to the ground but none the less dangerous) made ascensions in various parlour-trick effects. A lady in Montmartre rose weekly hanging from one while sitting on a settee, playing a violin.

Individual flights were hard to arrange, uncomfortable and expensive, and the luckless passenger was expected not only to help transport all the necessary equipment to and from the balloon but also to pay for any damages incurred. The going could be heavy on both counts. The huge, unsteerable balloons weighed up to 500 pounds and there was no predicting where or on what they might land.

And these were by no means the only problems. Alberto Santos-Dumont's first ascent included falling into a cloud of fog, being snagged in a giant oak, "shaken like a salad basket" and running out of ballast. Although he found the experience ecstatic ("The sun cast the shadow of the balloon on the dazzling whiteness of the clouds," he marvelled, "while our own profiles, magnified to a giant size, appeared in the centre of a triple rainbow!"), he realized that he

Balloons had not made much progress since Montgolfier.

would have to design a balloon of his own.

He engaged a tutor, to widen his scientific knowledge (without in any way endangering his belief that absolutely anything was possible), located a reasonably priced balloon-maker and set to it. The first thing he tackled was size. Originating the motto "Build lightness in" (inspired, perhaps, by his own spare frame), he created the little *Brazil*. It was made – against the advice of all the experts – of Japanese silk and bamboo and weighed just 44 pounds.

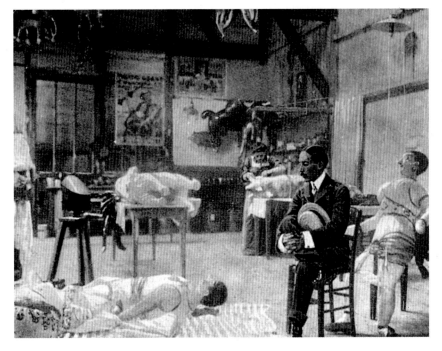

M. Lachambre also created balloon figures for circuses and fairs.

It was the smallest balloon ever made.

"The story that I carry it around in my valise," he admitted, "is true."

Although he was to go on to make twelve more balloons and several planes, this, his first creation since the Montgolfiers on the veranda, remained his favourite.

"My first balloon
the smallest
the most beautiful

Santos-Dumont insisted that his airships be made from yellow Japanese silk.

the only one named BRAZIL"

he wrote beneath its photo.

While sailing the little Brazil for enjoyment, he continued to experiment with motors, fuel, ballast, steering, rigidity and size, usually working against all the established principles of the day. Among his early advances, aside from the boldness of combining an internal combustion engine with explosive hydrogen, was the hanging of the gondola from ropes attached to the base of the balloon, instead of from the traditional heavy network of ropes covering its entire surface. Another was his development of a means of controlling ascent, descent and attitude through a series of movable weights, rather than solely by releasing gas or shedding ballast.

"Follow my directions and do not concern yourselves with their practicability," he ordered his workmen, a small group of

He wore his famous high collars even when ballooning.

former motor mechanics led by a solid *campagnard* named Albert Chapin who remained with him throughout his career. Grumbling at the start, they learned to act almost unquestioningly, hoisting a motor tricycle up into a tree to test its behaviour in the air or hanging a table and chairs on wires from his dining-room ceiling so that he could accustom himself to the feeling of eating in a balloon. (When the ceiling, inevitably, collapsed, they built a table and chairs six feet from the floor.)

No spartan, he considered lavish champagne lunches a normal part of his flights, as this description of a meal aloft amply illustrates:

"A joyous peal of bells mounted up to us. It was the noonday Angelus, ringing from some village belfry. I had brought up with us a substantial lunch of hard-boiled eggs, cold roast beef and chicken, cheese, ice cream, fruits and cakes. Champagne, coffee and Chartreuse. Nothing is more delicious than lunching like this above the clouds in a spherical balloon. No dining room can be so marvellous in its decoration. The sun sets the clouds in ebullition, making them throw up rainbow jets of frozen vapour like great sheaves of fireworks all around the table. Lovely white spangles of the most delicate ice formation scatter here and there by magic, while flakes of snow form

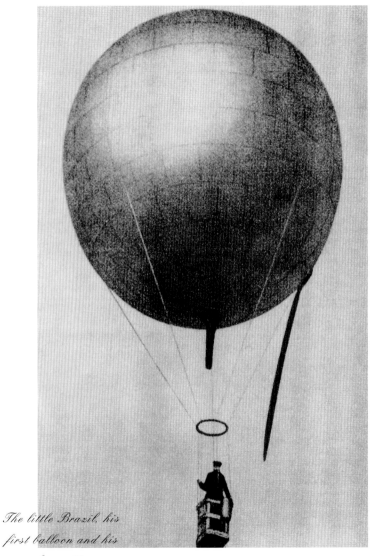

The little Brazil, his first balloon and his favourite.

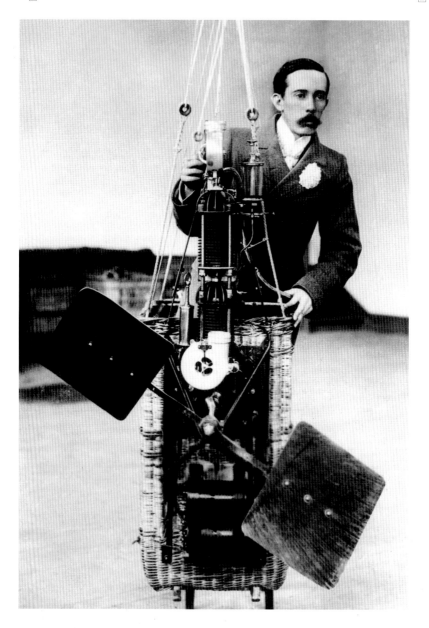

A diagram of the Brazil, which was so small he carried it in his valise.

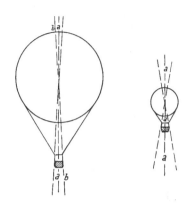

The engine and the propeller of the No. 1 were mounted on the basket.

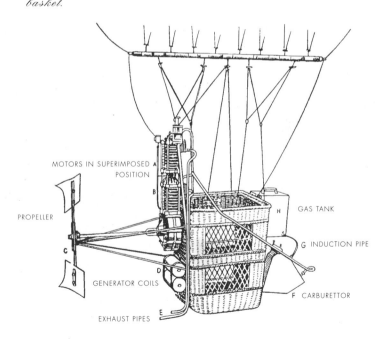

MOTORS IN SUPERIMPOSED POSITION

A

B

PROPELLER

C

D

GENERATOR COILS

E

EXHAUST PIPES

H GAS TANK

G INDUCTION PIPE

F CARBURETTOR

moment by moment out of nothingness, beneath our very eyes, and in our drinking glasses!"

His experiments did not go unnoticed.

Crowds of Parisians soon turned up, swinging their bowler hats on canes, to watch him guide-roping over their rooftops on the Champs Élysées (one citizen even complained he was being followed) and making test flights over the Bois de Boulogne or – prophetically – around the Eiffel Tower.

And, while the balloons he had found on arrival disappointed him, the automobiles – or motors, as they were then known – did not.

In addition to one of the first Peugeots (which he later took to Brazil, making it the first automobile ever in South America), he bought several motor tricycles (a kind of three-wheeler popular at the time, easy to handle and to race), a Panhard and an American "buggy". These could have been considered extravagances. Few, even among the wealthy, owned even one motor – they were so rare that the sight of them made people stop, holding up the traffic – but for Santos-Dumont they proved to be an excellent investment. Through them he found the power source he had been searching for.

Rejecting the idea of a steam engine ("weak in proportion to its weight and spitting red hot coals") and not dallying for a moment with the electric (which

An insufficient air pump caused the No. 1 to double up: crowds watched as it then crashed.

Though colleagues called his decision to use it suicide, Santos-Dumont felt he owed all his success to the combustible engine.

"though promising little danger has the capital ballooning defect of being the heaviest known engine"), he decided on the petroleum motor, such as that used in his De Dion cycle. It was lightweight and simple, and he knew from having taken it up into a tree that it vibrated less in the air than on the Paris cobblestones. While it was "still a delicate and capricious thing" and many advised against it – including his new friends, fellow daredevils all, who thought it "suicide" – he went ahead.

"My automobiling experience," he explained, "has stood me in good stead with my airships. I owe all my success to the combustible engine."

CHAPTER THREE

The High Life

"NOWADAYS I BUILD AIRSHIPS IN A LARGE WAY. I AM IN IT AS A KIND OF LIFE WORK."

For the next several years the yellow airships of Alberto Santos-Dumont were a familiar sight in the skies over Paris and the surrounding countryside as he pursued his unique designs, with varying degrees of success. Red banner flying with its Brazilian motto *"Por mares nunca d'antes navegados!"* ("Over seas yet unsailed!"), he could be seen dangling from a varying source of contraptions made of everything from taffeta to piano cord, and seated on (or in) anything from a basket to a bicycle saddle or even, at one point, simply a long pole.

His goal was a safe, affordable, steerable balloon and he was prepared to do anything to achieve it, including on one occasion asking a group of distinguished passengers to strip and throw their clothes over the side to lessen the weight.

This, fortunately, turned out not to be necessary but, so great was the esteem in which he was held that it would most

Known as "Little Santos", he had an "endearing oddness".
The works of Jules Verne still inspired him.

certainly have been forgiven, as were various other, often cost-ly, inconveniences his experiments caused. Known as "Little Santos", he was indulged by the Parisians as he had been by his family.

He was not the only aspiring aeronaut around – others, including Louis Blériot, an early racing chum and rival, was also at it – but he was their favourite. "He had," wrote one journalist, "an endearing oddness."

Slight (five foot six) and trim (110 pounds) with piercing, sometimes mischievous eyes, and a cleft chin, he was described as having "the agility of a cat, the sure feet of a climber [perhaps inherited from his father], the hands of an engineer, an extraordinary restless manner and an unshakable certainty."

The certainty resulted from his refusal to accept limitations – no one had ever told him anything couldn't be done – and his vision continued to be shaped by Jules Verne.

Although his methods were unorthodox and he could some-times, like Sherlock Holmes, be distant (throwing plane-shaped darts or picking bees off rhododendron bushes to study while people were trying to get his attention), he was admired by his colleagues for his courtesy, his workmanship and, by no means least, his courage.

"One only had to see him in a little wicker basket next to a motor going full out and vomiting flames beneath a balloon full of hydrogen gas to be convinced of that," declared Blériot.

And, all else aside, he got results.

"While we were busy losing time in unnecessary adjustments," wrote another ballooning rival, "he would be sawing up little pieces of bamboo and joining them together with bits of tin and string and in a few hours leave us open-mouthed with wonder and admiration."

He feared nothing but the number eight, which he regarded with an almost pathological terror.

Such derring-do was not without its dangers, of course.

He was stung by the bees. And storms, updraughts, ruptured air pumps and leaky valves (combined with his preference for trying things out to extensive paperwork) resulted in a frequent, and colourful, need for rescuing.

Descending too rapidly onto the grassy – but still hazardous – "ocean of greenery" of the Bois in his No. 1, he motioned to some quick-witted young boys who, exchanging their kite strings for his guide-rope, ran with it against the wind and managed to manœuvre him safely down.

Snagged in the tallest chestnut tree in the garden of Edmond de Rothschild in his No. 5, he enjoyed a luncheon sent up by a neighbour, Comtesse d'Eu, daughter of Pedro II, Emperor of Brazil, while the necessary repairs were being made.

Later, in the same balloon, hanging by his gondola from the roof of the Trocadero Hotel, he noted – while being cut down by firemen (to whom he later wrote a thank-you note) – that he had been "saved by the piano wires of Nice" which he had recently substituted for rope rigging.

Santos-Dumont awaits rescue after crashing into the Trocadero Hotel.

Although often at great risk, he managed to avoid serious injury. A bout of pneumonia was the only early threat to his health and, never losing the certainty of his childhood days on the plantation, he looked on each mishap as a lesson in design.

He often crashed into the Park of the Rothschilds where champagne was sent up during repairs.

"I attach little importance to my crashes," he said. "I believe in my star."

Trapped in the darkness by thunder and lightning, in a borrowed balloon, he felt "fierce joy" at being part of the storm.

Convinced that he faced death when his own cylindrical No. 1 began to break up, he felt "thrilled" at the thought of whom he might soon be meeting in the next world.

Someone was usually on hand to rescue the Sportsman of the Air wherever he unexpectedly descended.

"One is frightened," he once told an interviewer, "only when he still thinks he has a chance."

Nor did he let his demanding and frequently harrowing experiments interfere with his social life. Rising early to avoid the winds, he could put in a full morning's ballooning at the

A boater made a nice change from his usual panama.

Aero Club or sailing over Paris and still be at his regular table at Maxim's for lunch. As his celebrity grew he began to be written about in magazines and journals as much for his lifestyle as for his inventions.

Elegant, witty and smartly dressed, often sporting a panama hat, he began to set the pace in fashion as well. His high white collars (known as Santos-Dumont collars) and silk-lined opera cloak became all the rage.

His gold bracelet (holding the St Benedict's medal given him by the Comtesse d'Eu as protection against further crashes) and the thin, broad belt he wore beneath his waistcoat were noted and copied as were his boots (though not their lifts, which no one knew they held), his spats, his carnation button-hole and his driving costume of cap, goggles and knickerbockers.

A favourite at the numerous receptions and balls of the day, he also enjoyed playing host, and many of the wealthy and titled, including the Rothschilds, into whose lakes and rose gardens he so often crashed, had been guests in his luxurious house at 5 rue Washington the evening before.

Prominent among them were Antonio Predo, the Brazilian Ambassador, and George Goursat, the famed caricaturist known as SEM, whose original drawings can still be seen on the walls of Maxim's and London's Pont de la Tour and whose distinctive work is still copied, and collected, as emblematic of the era. He and Santos-Dumont, in their matching boaters and fashionably turned-up trousers, were

MAXIM'S

The famed caricaturist SEM, *a close and lasting friend.*

SEM

An invitation to dine from Alberto Santos-Dumont to Louis Cartier.

RIGHT

The two friends practise dining aloft.

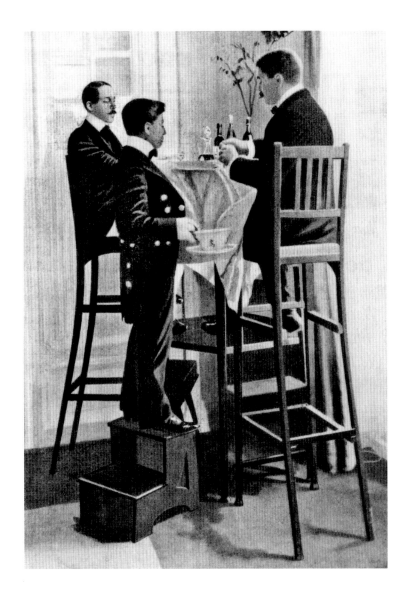

very much men about town, attending concerts and gala openings, with SEM often sketching his more illustrious companion. Their friendship was a close one and, based on more than café society, outlasted most others of that time.

Another frequent visitor was the jeweller Louis Cartier, who was to feature so prominently in his life and the resurgence of his name and reputation. From one of their meetings came the inspiration for the Santos, the first wristwatch, and still Cartier's best-selling design.

Cartier also created other jewellery for Santos-Dumont, notably a ruby-studded piece engraved to Belle de Neuilly, mysteriously described as the balloonist's mistress. No further reference was ever made to her but this and a later – and very public – infatuation with an aristocratic Cuban beauty are the only recorded hints of any involvement he may have had with women, who were otherwise absent from his life.

He never married and had close male friends, but this was not unusual for his time and there is no evidence of any gay relationship, either. Indeed, many, if not most, gay men of his class, including Oscar Wilde, used marriage as a cover. (A number went even further, spending fortunes in setting up supposed mistresses – tolerated by wives of the day – in expensive houses which they then visited to be with the male lovers they preferred.)

The fact that he had not married was no more proof that he was gay than the posthumous judgement made by researchers in a more innocent (but not that long ago) time that his "m e c h a n i c a l

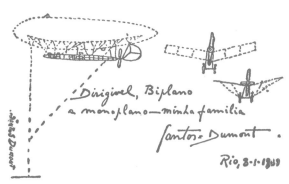

"Dirigible, biplane and monoplane – my family" wrote Santos-Dumont on an autograph for a fan.

interests" and "indifference to the arts" proved "utterly" that he was not.

He may have been gay and, wisely, discreet about it.

He may, as his friend Goursat (who later did marry) said of him, have been just "extremely shy and reserved".

Or, an obsessed visionary, he may merely have wanted to travel light in all senses of the word. While signing an autograph for an admirer he once sketched an airship, a monoplane and a biplane, writing beneath them "my family".

In any case, in Alberto Santos-Dumont's version of the high life, romance did not seem to be a top priority.

CHAPTER FOUR

Conqueror of the Air

"I DID NOT YET KNOW MY EXACT TIME. I CRIED, 'HAVE I WON?'"

O n 19th October 1901, the sportsman of the air, as he liked to call himself, had his first big success, winning the prestigious Deutsch Prize for a half-hour controlled flight around the Eiffel Tower in his cigar-shaped No. 6. The Deutsch Prize was offered by the Aero Club and donated by its president, the wealthy petroleum magnate Henri Deutsch de la Muerthe, who shared Santos-Dumont's interests in both petroleum (the base of his family fortune) and Icarus (about whom he had written an opera). It had been preceded by an Encouragement Prize, awarded to the balloonist who seemed to be trying hardest.

This, of course, had also been won by Santos-Dumont, who had promptly returned it to found another prize: for anyone, except himself, who made the flight – untimed. This award, although never actually

A little knot of onlookers was always on hand.

Santos-Dumont designed and paid for all his machines himself. He held no patents and took no money for any of his inventions.

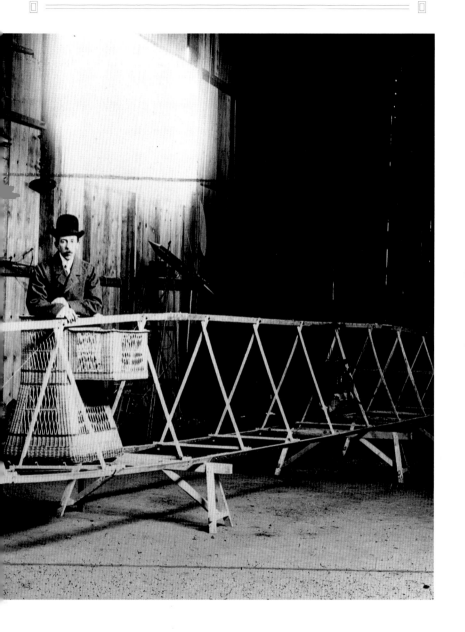

collected, left Baron Deutsch de la Muerthe slightly miffed, since it did away with his proposed condition of a half-hour time limit, and may have influenced some of his later, less generous actions.

In any case, Santos-Dumont won.

It was his third try, several others had been killed in the attempt and, midway, he had to climb out of the basket and scramble across the girders to adjust the misfiring engine. A thousand feet below, Paris held its breath, then cheered. The streets were so crowded that men were falling into the river from the parapets of bridges.

"Have I won? Have I won?" he asked as he landed and in

answer the crowd lifted him from the gondola and carried him on their shoulders to the waiting Aero Club judges.

That night, recounting the adventure at a gala celebration at Maxim's, he described the problem of keeping time while flying.

All is suspense as the Aero Club judges ponder the time.

Rounding the Eiffel Tower in his No. 6 to win the Deutsch Prize.

The business of steering alone (not to mention extinguishing flames with his panama hat or scrambling among the girders) left no hand free to check his pocket watch, which was, of course, why he had had to ask the crowd if he had won. He did not, as he explained, know his time.

The inventive Cartier listened carefully and soon came up with the solution: a watch that could be strapped around the wearer's wrist. It was stylish and sporty like its inspiration and, like so many other chic (and some not so chic) items of the time, named after him as well. Medallions, charms, postcards, caricatures and *bibelots* – which translates as trinkets, curios, knick-

Receiving a tribute from an admirer.

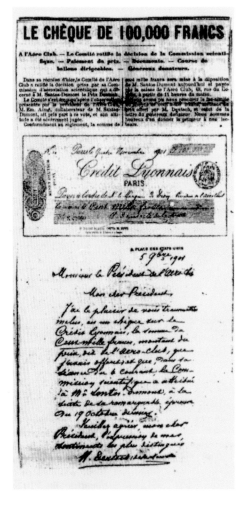

The prize was given away – half to his assistants, half to recover tools pawned by Parisian workmen.

knacks — were created and coined in his honour by the thousand. Except for very rare instances, few remain. (A pendant dedicated to Louis Blériot, on the other hand, was for sale in the summer of 1997 for £2,700, in the Burlington Arcade off London's Piccadilly.)

Santos-Dumont had become an international celebrity. At a time when men were still trying to build artificial wings, he had conquered the air.

Honours and invitations poured in from all over the world. Pioneers such as Guglielmo Marconi and Thomas Edison (who congratulated him on his choice of the petroleum engine) sent their photographs and welcomed him to their ranks.

France was ecstatic at now having grabbed the lead in science as well as everything else. Other countries such as Germany (where Count von Zeppelin's experiments had yet to come to fruition) and the United States (where no one had yet even heard of the Wright brothers) were worried.

The prize of 100,000 francs he gave away, half to his assistants, and half to recover the tools pawned by Parisian workmen. (An additional 125,000 francs, awarded him by a proud and grateful Brazil, he kept for himself and for further experiments.)

Of all the congratulatory letters he received, the one that pleased him most was from Pedro, a former playmate, reminding him of the days on the plantation and their games of "Pigeon Flies!".

"Man flies! old fellow!" cheered Pedro. "You were right to raise your finger. You have just proved it by flying round the Eiffel Tower and M. Deutsch has paid the forfeit in your stead.

"They play the old game more than ever at home since your flight," he added, "but the name has been changed. Now they call it 'Man Flies!' and he who does not raise his finger must pay the forfeit."

Closer to what was now home, new friends beckoned. England, too, was inspired by this sportsman of the air. Led by motoring pioneers Frank Hedges Butler (a wealthy wine merchant) and the Hon. C. S. Rolls (who had yet to team up with

An artist's fantasy panorama – Paris as it must have looked from the No. 6.

Mr Royce), the Aero Club of Britain was founded and Alberto Santos-Dumont invited to become its first guest of honour.

So it was that on 23rd November 1901,

a nervous Santos-Dumont, whose fear of public speaking had turned out to be as great as his terror of the number eight ("He would rather face a firing squad," said Goursat), addressed the leaders of the British scientific world at a gala banquet at London's Carlton Hotel – in English.

Among those attending were the heads of the Royal Astronomical Society, the Royal Geographical Society, and the Society of Electrical Engineers, Sir C. Champion de Crespigny, for the Royal Navy, the editors of twenty-one newspapers including *The Times*, the *New York Herald*, *Le Matin*, *O Globo* and *Lady's Pictorial*, and, fittingly, a Mr Bird and a Mr Pigeon.

Over the cigars, two ironic toasts were made: one to the future of ballooning in war, another to the fact that though many prominent names would be forgotten, one that would always be remembered would be that of Alberto Santos-Dumont.

All was not adulation, though.

"England is no longer an island," complained Lord Northcliffe of the *Daily Mail*, jumbling his images somewhat. "There will be no more sleeping safely behind the wooden walls of England with the Channel our safety moat. It means the aerial chariots of a foe descending on British soil."

None of these predictions seemed worth thinking further about at the time, and Santos-Dumont spent the rest of the week being entertained at such places as the Royal Thames Yacht Club, the Reform Club, the Brazilian Embassy and Aldershot — where he ascended with Mrs T. B. Brown and other ladies in a captive

"The more I can escape the feeling of gravity, the better I feel."

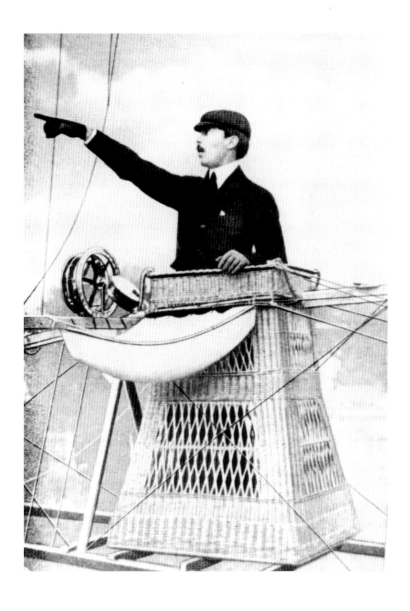

balloon called Empire, made of gold-beater's skin. (This was actually the lining of cows' intestines, so-called because it was used to hold gold as it was hammered into gold leaf.) At a dinner given for him by Mr Paris Singer a model of the No. 6, four feet long and three feet high, was constructed of violets and lilies of the valley.

Promising to return in, he hoped, more clement weather, he went back to France where the Grande Exposition was uniting

the world in peace, and his celebrity continued at fever pitch. He spent the next month enjoying it, accepting tributes, meeting with anyone who wanted to know more about his ideas, and, as always, dining with his friends at Maxim's.

Social life alone, however, never satisfied him and just as he began to get restless, an invitation came from Prince Albert of Monaco. The grandfather of the present Prince and creator of the famous Oceanographic Museum ("himself a man of science celebrated for his personal investigations") offered to finance the building of a hangar and base for airship flights over the sea.

Santos — Dumont

'THE CONQUEROR OF THE AIR'

AFTER Bartolomeu Lourenço de Gusmão, who was to go down to history as the "flying priest," thus immortalising a nickname which is not without its humorous connotations, there is no other name so famous in the long list of pioneering Brazilians in the conquest of the air than that of Alberto Santos Dumont.

Born in Minas Gerais on 20th July, 1873, he grew up on the vast coffee plantations in São Paulo, belonging to his father, an engineer who took his degree in the Central School in Paris. From early childhood his vivid imagination was fired by the stories of Jules Verne and he showed a deep interest in machinery of every kind. In his book 'In the Air,' in which he recapitulates the whole of the first phase of his struggle for the conquest of the air, he says of himself: "When I was seven years old I already had permission to drive the large wheel locomotive used in our plantation for field work. At twelve years of age, they used to let me substitute the engine driver in the Baldwin locomotive which pulled the coffee transportation trains

CHAPTER FIVE

Darting Out on the Mediterranean

"THE SITUATION PROMISED TO BE IDEAL."

Enticed by the challenge of ballooning over water, as well as by the chance to escape another raw Paris winter (the temperature in Brazil rarely drops beneath 22 degrees Celsius all year round), Santos-Dumont spent the Christmas and New Year's holidays organizing his departure.

The situation, he felt, promised to be ideal. The little bay of Monaco, sheltered against the wind and cold, would make a well-protected manœuvring ground from which he might, as he imagined it, "dart out on the Mediterranean".

Sending ahead first his suggestions for the construction of the balloon house, his mechanic Chapin and the No. 6, and his brand new Mercedes and chauffeur, he (and his extensive wardrobe) then set off happily for the Riviera on what was the original model for today's Orient Express.

Installed in his own suite at the villa of the Duc de Dino, he spent several months living roughly the sort of life he had in Paris: glittering evenings followed by days filled with hair-raising and death-defying experiments.

Monte Carlo was at its peak; its hotels and yacht basin lured the wealthy and titled from all over the world, including the American millionaire James Gordon Bennett, founder of the *Herald Tribune*

The hangar in Monaco, though large, was not ideal.

and a keen amateur balloonist, who soon became a great fan and supporter.

The stylish and daring adventurer became the latest attraction along the Côte d'Azur, rivalling the Casino, theatre, formal gardens, opera, royalty spotting, band concerts and admiring and/or envying one another's ensembles at galas and balls. Even Napoleon's widow, Princess Eugénie, broke her self-imposed exile to come and meet this marvel.

Launching the No. 6 was not easy.

His progress was avidly watched from yachts and hotel balconies as, in less than a month, he oversaw the building of a large aerodrome and hydrogen-generating plant, setting another world record in the process, this time for the heaviest sliding doors ever made. Although it was predicted that, at over half a ton each, they would never open, and even his own workmen were inclined to doubt their feasibility, they were so delicately balanced that two young children (the princes Ruspoli, aged eight and ten) managed them at the christening ceremony with no effort at all.

Getting the prizewinning No. 6 into the air was a much more difficult job. The problem was the hangar, or balloon house, as he preferred to call it. Instead of being over the sea, as the Prince's promise had implied, it was across the boulevard, tram tracks, promenade and sea wall. Each of these, plus a fifteen-foot drop to the sand, had to be negotiated before a take-off could be attempted. Although it was finally managed, amid great cheers, Santos-Dumont, jaunty in dark suit, high collar and motoring cap (oddly, he never created a special ballooning costume), was nearly thrown into the water in the process.

Getting the balloon (and him) back down represented even more of a challenge. As there was no landing platform, the airship had to glide directly into the huge hangar, rather like thread into a needle.

"I steered boldly," he later recounted, "and was able to make a sensational entry without damage—and without aid!" But the slightest gust of wind could have sent the airship crashing into the wall, the boulevard, or, as he himself, "the interested party", pointed out, the sharp-cornered buildings on either side.

Something had to be done. The Prince offered to tear down the wall but in the end, ever polite, Santos-Dumont demurred, settling for a landing stage to be built above the sea.

But it was a harbinger of things to come. Darting out on the Mediterranean proved not to be as much of a delight as he had imagined.

Newly-sprayed with yellow lacquer, the No. 6, to which many attributed the kind of intelligence currently ascribed to computers, found itself facing strange new challenges: battling frequent rain, tricky sea winds and the over-enthusiasm of the well-meaning but inexperienced Prince Albert who, following below in the Royal Yacht *Princess Alice*, sent up a dangerous blast of smoke and sparks, and entangled himself in the guide-rope (badly injuring his right arm) – all in the same afternoon.

Although Santos-Dumont had some happy times, flying along the coast, followed by yachts at sea (and, memorably, waved at from one by a "graceful female figure" with a red *foulard*) and elegant motor cars on land, the whole exercise was something of a let-down.

While he enjoyed the absence of such hazards as the roofs and chimney pots of Paris, he missed the comfort and company of its crowds, always there to encourage and, if necessary, rescue (and send up lunch to) him as well.

Here, though surrounded by offers of help while on the ground, he often found himself conducting his actual experiments "isolated" as one onlooker described it "over the immense sea".

Many volunteered "to be on the spot" in case of accident, among them James Gordon Bennett and the Prince in their yachts and Mr Clarence Dinsmore and Mr Isidore Kahenstein, two local characters, in (respectively) their 40hp Mors and 30hp Panhard. But it was not always possible to keep up with him, carried away as he so often was by swift winds and impulse.

The perfect spot for a bit of darting out on the Mediterranean.

"This kind of protection must not be counted on overmuch by airship captains," he decided somewhat sadly.

The truth was, he was lonely. And he admitted it. "The loneliness in which I found myself in the middle of this first extended flight up the Mediterranean shore was not part of the programme," he later wrote.

Any decision about whether to remain was taken out of his hands when, on Valentine's Day, owing to an imperfect inflation, the beautiful No. 6 reared up, disarranging the rigging, which gave way and caught in the propeller, and, surrounded by a little flotilla of boats, it sank.

Like a true captain, Santos-Dumont refused to leave his ship, trying until the last moment to have it towed to shore. Only when he was up to his neck in water did he let himself be

As the Prince and other rescuers await, the No. 6 prepares to sink.

hauled to the safety of the *Princess Alice*, and even then, dripping wet, he directed the salvage of its remains, cheered on by thousands of waving handkerchiefs from shore and the spelling out of his name in white stones along the sand.

Yet another great banquet was given in his honour that night and a fund begun for the rebuilding, but he decided to return to Paris instead. "It is," he declared, "the best place for airships."

He bundled up the remains of the No. 6, and he and his little caravan headed home.

Reflecting on his experiences Monégasque, he decided several things.

He wasn't sorry to have accepted the Prince's offer over other similar ones made at the time. After the stress surrounding the Deutsch Prize, he had enjoyed "amusing" himself with his airship and making observations of use to him alone, not having to prove anything to anyone else. And, while he was not averse to an occasional endorsement for something like, perhaps, a liqueur, such as Bénédictine, he did not want, as these others (in Britain and, interestingly enough, the United States) decidedly did, actual advertising slogans on the sides of his airships.

As to the crash, although it had been blamed on them at the time, it was not entirely the fault of Chapin and the other mechanics. There hadn't been enough space in which to prepare the No. 6 properly, and the transition from the cool hangar into the hot sun had speeded up its inflation. He was, though, going to have to retrain his mechanics in future if he began new experiments. And, meanwhile, he would have to "groom" the airships with his own hands as he had at the start.

Additionally, and with astonishment, he realized that it was on this trip that he had, "unperceived", come closest to disaster.

It had been the day of his most successful flight when the Prince had been hit by the guide-rope as his yacht, smokestack spewing sparks, was trying to tow him in.

"Any one of those red-hot sparks," he recalled, "might have, ascending, burnt a hole in my balloon, set fire to the hydrogen, and blown balloon and myself to atoms!"

CHAPTER SIX

The Racer

"I AM PROUD TO SAY THAT I HOLD NO PATENTS."

On his arrival in Paris, he found that his fellow balloonists had not been idle in his absence. While they had cheered his achievement, it had also inspired them to try to top it, often through use – or misuse – of his own methods, on which he was famed for never taking out patents.

That old colleagues such as Blériot and the Farman brothers (two young Englishmen who had grown up in France) were hard at work on new projects was no surprise. But now they were by no means the only ones.

His style and courage had made it all look easier than it actually was, and there was a kind of frenzied infatuation with aeronautics afoot. Wild predictions were being made, and sketched in the papers, about balloon metros, balloon gendarmes and even a

Although the new hangar was small, he envisaged it as something like Heathrow.

Transatlantic airship flights were eagerly anticipated.

balloon Grand Prix, to be watched by VIPs from balloon boxes. Santos-Dumont was not entirely innocent in this regard, talking dreamily of guide-roping around the North Pole and the advent of great airship liners. (His brother, Henrique, on the other hand, couldn't get anyone to pay much attention to his dream — a tunnel beneath the English Channel.)

The inevitable result of such speculation was that absolute beginners were now also taking to the air, more often than not with disastrous results.

The most spectacular of these was the crash of the Pax, the creation of a fellow Brazilian, Augusto Severo, who had arrived with a dangerously premature plan for a regular transatlantic service for more than 100 passengers.

Too lightly weighted, it shot up into the air immediately on its first trial, disappeared from view (thanks to the panic of its creator and his mechanic who were last seen throwing out ballast instead of valving gas), exploded and landed in pieces, which were cut up and sold as souvenirs to the horrified crowd. (This had particularly unpleasant associations for Santos-Dumont. Not only was Severo the second Brazilian balloonist to

In the future, even commuters would take to the air.

Even an airship Grand Prix was predicted.

come to a bad end – one Laurenco De Gusmao, known as the flying priest, had been put to death as a sorcerer by the Inquisition in 1709 – but the doomed airship had been built by Monsieur Lachambre, who had built his own beloved Brazil.)

Issuing what warnings he could and, faced with the traditional problem of the early achiever – what to do next – he tried not to feel guilty as gondolas fell from the sky and enormous white elephants refused to leave the ground.

He started work on another airship, the No. 7, which, though he never raced, he called, perhaps in an unconscious nod to his new situation, the Racer. Nothing much came of it; he complained, ironically, that he could find no one to race with him. In a way, his very success seemed to hamper him.

It kept the invitations coming, though, and, somewhat at loose ends, he accepted a great many of them.

He made a triumphant visit to Brazil, where he was hailed as a hero and presented with a model of the No. 6 in gold and diamonds.

He returned to England (where the weather had not improved), oversaw the reconstruction of the No. 6 for display at the Crystal Palace, visited his niece and his favourite nephew, Jorge, who were at school in Tavistock, and challenged a member of the Savage Club to a duel at the Ritz.

The duel averted, he carried on.

He reviewed the entire French army from above on Bastille Day (and startled the President of France by firing a 21-pistol salute from his airship). He considered – and, happily, rejected – an expedition to the North Pole. He went to St Louis, Missouri, to advise the planners of the 1904 Exposition on the

The French army were startled when "Little Santos" fired a 21-pistol salute over their heads.

organization of airship races, stopped in New York to meet with Thomas Edison, and advised both Teddy Roosevelt and Admiral Dewey – which may be why, to this day, the United States continues to be involved with airship development.

Cheering crowds and press coverage were as phenomenal in America as they still were everywhere else and, while no one had yet heard much of the Wright brothers, it seems hard to believe that the Wright brothers could have missed hearing about Santos-Dumont, a supposition later cited as evidence against their claim of a flight in 1903.

If they had indeed flown by then, went the argument, why didn't they enter the competition at the Exposition in 1904 which was not only widely publicized (and nearby) but offering a large amount of money in prizes? Especially since, unlike Santos-Dumont, who never asked for (or received) money for any of his inventions, they later travelled all the way to France to try to sell theirs to the French government?

This was all in the future, however, and for the moment Santos-Dumont had other things on his mind.

The punters weren't the only ones who had taken up ballooning. Large business firms were now investigating new forms of flight, and setting aside huge sums of money for the job.

Even his former patron, Henri Deutsch de la Muerthe, was now a competitor. No longer offering prizes to others, he had founded his own company, Astra Airships, constructing an enormous hangar on the Aero Club grounds directly opposite Santos-Dumont's.

As the interest of big business grew, so did the idea of the use of balloons in war.

The 21-pistol salute fired over the Grand Review had hardly gone unnoticed and both the press and the Minister of War were speculating on further possibilities.

A dirigible with a keel made of steel tubing constructed by two wealthy French sugar refiners, Paul and Pierre Lebaudy, had also attracted the attention of the French military.

Count von Zeppelin was waiting in the wings.

The Aerial Yachtsman was a pioneer of all sorts of transportation.

A traffic jam aloft and (left) returning from the opera in the year 2000.

The days of the dashing dedicated amateurs, the sportsmen of the air, were numbered.

These new developments troubled him.

His response to the increased (and increasingly powerful) competition was to consolidate his efforts.

Removing himself from the intrigue of the Aero Club, he built a new hangar – this time preferring to call it an airship station – in Neuilly, between the Bois and the Seine. Although

Santos-Dumont was not alone in taking Icarus as his symbol. Many other men of the air did the same.

nowhere near so grand as the one in Monaco — and really only a large red and white striped canvas tent — it could hold seven fully inflated airships, and he thought of it as "the first of the airship stations the future must produce for itself".

These he envisaged as something like Heathrow, with cars to pull the airships in and out, observation towers, telegraph equipment ("able to communicate with distant goals and, perhaps, even with airships in motion"), workshops, generating plants and "sleeping rooms for experimenters who desire to make an early start and profit by the calm of the dawn". Everything but the duty-free shopping.

His feelings towards the growing militarism, however, remained unresolved.

As an idealist, he clung to his original plan, which had always been that airships would contribute to peace by making the world smaller. As a scientist, he couldn't help but see the many uses they could have in war. Each new discovery seemed to make such a departure more inevitable as, when soaring over the Mediterranean, charting winds and doing guide-rope experiments, presumably concerned about nothing but the "shoot the chutes" (or *montagnes russes*) – the unpredictable air currents of changing altitudes – he had kept realizing how easy it would have been also to be tracking submarines.

But the military application of his inventions still seemed a long way off. Ever courteous, he pledged his fleet of airships to the government should they need them (with the naïve condition that they were never to be used against either North or South America) and put the question to the back of his mind.

War, for the moment, looked unlikely.

What was at risk was his supremacy in the air.

CHAPTER SEVEN

The Aerial Promenader

"I TAKE MY PLEASURE IN THE BOIS IN MY SMALL No.9."

The next two years encompassed a great deal of activity, although a reference made to them as a "rest period", which many found amusing, was not altogether inaccurate. He constructed his beloved No. 9, La Baladeuse (the Stroller), which became known as as his little runabout, and used it as one might nowadays a favoured sports car or convertible for jaunts over the countryside and, extensively, landings for lunch at La Cascade, a favourite restaurant at the edge of the Bois de Boulogne.

His crashes into smart hotels and secluded residences had, in a way, been replaced by leisurely landings. He dabbled, one might say, in landings.

No longer content merely to guide-rope along the Champs Élysées, he now touched down before his corner house on rue Washington where his valet, Charles, had coffee waiting. Certain, like H. G. Wells, that balloons would one day be as

Santos-Dumont used his No. 9 for aerial "strolls", and luncheon landings at one of his favourite restaurants, La Cascade.

common as cars, he planned an ornamental landing stage extending out from his curved bay window.

"I thought of the time, sure to come," he wrote, recalling the event, "when the owners of handy little airships will not be obliged to land in the street but will have their own guide-ropes caught by their domestics, on their own roof gardens."

On another occasion, much to the surprise of a boulevardier trying to enjoy an iced orangeade, he landed on the terrace of a café on the rue Bois de Boulogne, downed an apéritif, made his apologies, and was off again.

Coincidentally (and unfortunately), the very same boulevardier was similarly surprised in several other locations during the next few days — driving, strolling and heading home late at night after sampling new drinks of American origin — and eventually complained to the police.

While he was a bit of a show-off, and his delight in amazing people occasionally backfired, Santos-Dumont always maintained an unerring politeness and sense of propriety — even to monuments. Sailing by the Arc de Triomphe one day, he felt tempted to guide-rope beneath it but restrained himself, he later confessed, because he did not feel "worthy".

Sometimes now, though never on trials, he even took up an

The crowd watched as he landed the No. 9 at his house, where Charles, the valet, had coffee waiting.

unlikely passenger just for fun.

Descending amid a children's fête at Bagatelle, in the Bois, on 26th June 1903, he asked, "Does any little boy want to go up?" From amongst the many

Though a bit of a show-off, he was always polite.

eager volunteers he chose the closest, the son of an American diplomat.

"Are you not afraid?" he asked the seven-year-old as the airship rose. "Not a bit!" replied the doughty little passenger, to the delight of his captain, who declared, "The boy will surely make an airship captain, if he gives his mind to it."

But the boy, Clarkson Potter, did not give his mind to it, becoming instead the founder of the American publishing house that bears his name today.

While, in those pre-feminist days, it would have been heresy to ask if any little girl wanted to go up, it was only a few days later that a big and "very beautiful" girl made such an unusual request herself.

Her name was Aida d'Acosta and she was a famous New York society beauty from an old Cuban family. After several visits to his hangar at Neuilly (a must for *tout Paris*), she confessed what he referred to as "an extraordinary desire".

At first he thought she wanted merely to ascend with him and was mightily impressed by her courage. "Mademoiselle, I thank you for the confidence!" he told her. When he realized she wanted to go up alone – "and navigate free, as you do," she insisted – he was astonished.

And everyone else was astonished when, after three lessons, he allowed her to. On 29th July 1903, he noted, lest it go unrecorded, "The first woman to mount, accompanied or unaccompanied, in any airship, actually mounted alone and drove the No. 9 free from all human contact with its guide-rope for a distance of considerably over a half-mile (from Neuilly to Bagatelle)," an event he considered both "memorable in the records of dirigible ballooning" and "worthy of preservation in the annals of aerial navigation".

His activities were avidly chronicled by the papers of the day.

Even more memorable and worthy of preservation in the annals of aerial navigation was that this was the first and only time that anyone, including his closest design associates or experienced balloonists, EVER flew a Santos-Dumont airship other than Santos-Dumont himself.

She was a famous Cuban beauty and the only other person ever to fly a Santos-Dumont airship. Her picture dominated his desk for years after.

"I think the simple fact that I consented speaks eloquently in favour of my confidence in the No. 9," he later explained, but it was "the intrepid girl navigator" herself – in a huge and dashing hat – and not the No. 9 whose photograph dominated his desk for years afterwards.

He planned an ornamental landing stage extending out from the window of his house.

Not all his adventures in the snub-nosed No. 9 were quite so gay. The dreaded *retour de flamme* finally occurred. As the balloon floated lazily over the Seine, sparks shot back from the engine and started a fire. But even this did not diminish his fondness for the smallest of his dirigibles, which remained a favourite. "I promptly extinguished the flame with my panama hat," he reported, "and continued without other incident."

"I take pleasure in the Bois in my small No. 9," he said of

those days, "navigating through the cool air of the delicious dawn."

His descriptions of the delight he took in ballooning, often focusing on the air itself, have a gentle beauty:

"All is pure. There is no smell in the air."

"The balloon seems to stand still in the air while the earth flies past underneath."

"The basket rocked prettily, beneath the balloon, which the mild, fresh air was caressing."

As the general ballooning frenzy seemed to have abated slightly – the populace perhaps wearying of watching men and machinery hurtling down from the skies – his own urgency also lessened.

He constructed several more airships (though never, of course, a No. 8) but his heart did not seem really in them. His No. 10, the Omnibus, with little baskets for passengers; No. 11, a glider tested by being towed behind a motor boat; and Nos. 12, a helicopter for which no engine could be found, and 13, a kind of aerial yacht with its own heating plant, intended to stay aloft for a week), came and went without much fuss.

He seemed happier in his role as aerial promenader, luncheon party host and

Somewhere along the way he tried to leave airships behind him.

The song was in his honour but the sketch did not resemble him.

author, writing (with quill pen on his favourite stationery) *Dans l'air*. Published in 1904, at the height of his first fame, it perfectly reflects his jaunty charm. It includes, in the opening chapter, an explanation – at once amusing and accurate – of the manufacture of coffee.

"The berries of black coffee are red when they are green," he explained, adding, "Though it may complicate the statement, they look like cherries."

To some he may have appeared a restless mind with no direction at this point, but it was not that simple. His aim had been flight, not fame, and he had achieved it – and been made a Chevalier in the French Légion d'honneur because of it.

Perhaps his meetings with great men of the time, who accepted him as their equal, and his somewhat over-the-top reception in Brazil – where he had also been serenaded nightly beneath his windows with a song composed in his honour – had left him feeling fulfilled, if not sated.

He could have stopped at that point,

Curiously, though always stylish, he never had a special ballooning costume.

as did his book and, at thirty-one, counted himself a success. But perhaps, and even he himself may not have realized it, he was merely biding his time, while a new challenge germinated.

Goursat, who probably knew him better than anyone, reported coming upon him one day shooting plane-shaped darts with an "intense" expression on his face.

In any case, somewhere along the way he decided to leave airships behind him and concentrate instead on heavier-than-air machines.

He had no idea of the significance of Wilbur and Orville Wright's launch of their plane, the Flyer, into a controlled 59-second flight at Kitty Hawk.

And, aside from the five spectators, mostly coastguardsmen who happened to be on hand, neither did anyone else. They were still unknown to the world at large.

The coast seemed clear.

Crowds often gathered to watch the trials at Bagatelle.

IN DEFERENCE TO ALBERTO SANTOS-
DUMONT'S AVERSION TO THE NUMBER,
THERE WILL BE NO CHAPTER EIGHT.

CHAPTER NINE

A Monstrous Hybrid

"BIRDS DO THE SAME THING."

By 1905, Alberto Santos-Dumont was ready at last to tackle something new. He joined the growing ranks of dreamers, mechanics, balloonists and businessmen who had also decided that the future lay in heavier-than-air machines. Although it took them quite a while to realize it.

While his fame had in no way diminished, he was fixed in other people's minds as a balloonist, an aerial sportsman enjoying, as he himself put it at the end of *Dans l'air,* the "reward" of his early success. When he began testing a strange, box-kite-like craft

Even when experimenting with heavier-than-air machines, he found it hard to abandon his airships.

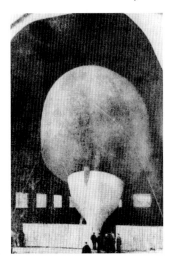

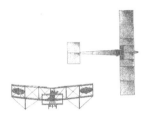

A sketch of the 14-bis.

hung from an airship and pulled by a donkey, they were more amused than concerned.

Dignified – and determined – as ever, he persisted, ignoring the jibes of the crowds and the witticisms of his friends. If, as he had come to believe, aerodynamics were the way forward, he intended to be the first to lead that way.

His yellow airships trailing their red banners were less frequently seen over the rooftops of Paris. His take-off cry of "Let go all!" was less often heard. Adding a young architect, Gabriel Voisin, to his team, he almost abandoned his social life, replacing evenings at Maxim's with talking and sketching late into the night.

His old mechanic, Chapin, remained with him to the end.

While Voisin, who had declared that he wanted to "consecrate his life to aviation", was experienced and had designed, and even piloted, airships for other balloonists, the initial results of his work with Santos-Dumont did not augur well.

Since he had moved it from Aero Club grounds, his hangar was no

The cartoonists now portrayed aviators and planes falling on hapless pedestrians.

longer under the constant scrutiny of his colleagues but a little knot of onlookers was always on hand – including reporters from James Gordon Bennett's *Herald Tribune*. To them, the

The strange, box-like contraption pulled by a donkey caused more amusement than concern.

tests seemed "a shambles of shouted orders, roaring engine, scampering handlers and the small inventor yelling at the top of his voice".

Of the many machines being made, mostly along the lines of gliders fitted with internal combustion engines, Santos-Dumont's looked the least likely even to get off the ground. And even the more likely-looking ones didn't look too promising either.

A glider commissioned by Blériot, which Santos-Dumont couldn't help noting had many of the features of his own No. 11, including being pulled by motor boat, crashed dramatically into the Seine on one of its early attempts.

Less prominent pioneers weren't doing much better and

cartoons predicting airship traffic jams, metro stations, transatlantic crossings and trips to the opera were being replaced by ones depicting pedestrians cowering beneath opened umbrellas in an attempt to avoid falling planes, aviators and other debris.

Santos-Dumont's contrivance comprised a series of box-kites held together with pine struts and piano wire, with the tail in front and propeller behind. As with the No. 6 in Monaco, the problem for him lay not in falling from the sky, but rather in getting up there in the first place.

Initially he hung his contraption from a rope on a pulley to be powered by the donkey. When this failed, he attached it to an airship built for the purpose, the No. 14. He was reluctant to abandon his airships, which had taught him so much about manœuvring currents that were useful. "Birds do the same

The ungainly combination of airship and box-kites was dismissed as "a monstrous hybrid".

A stubborn Santos-Dumont was determined that his strange new craft would conquer the air once again.

thing," he explained.

But the kite section kept jumping ahead and leaving the airship behind. Finally, he broke up what was being referred to as "a monstrous hybrid", put in a more powerful engine, called the result the 14-bis (i.e., 14 again or, for the less romantic, 14a) and felt ready to proceed.

It was, of course, more complicated, more dangerous and

less amusing at the time than it seems in retrospect. And considerably less respectable.

No lesser a pundit than the Editor of *The Times* had written at the beginning of that year: "All attempts at artificial aviation are not only dangerous to life but doomed to failure from an engineering standpoint."

Sir Stanley Mosely, meanwhile, opined: "It is complete nonsense to believe flying machines will ever work."

With his usual obstinacy, and spurred on, once again, by the offer of a prize – or in this case, two prizes – Le Petit Santos, as he had in the past, ignored all obstacles and forged, seemingly unwisely, ahead.

An excited crowd escorted their hero to Bagatelle, where he had once been rescued by the kite-flying boys.

CHAPTER TEN

Nothing Was Ever the Same Again

"19TH OCTOBER 1901 AND 12TH NOVEMBER 1906 WERE THE HAPPIEST DAYS OF MY LIFE."

Even his closest friends were astonished — and not a little worried — when he entered his name for a somewhat incestuous award known as the Archdeacon Prize. Actually it was two prizes: 1,500 francs from the Aero Club for the first heavier-than-air flight of 100 metres, topped up by more for less (another 3,000 francs for just 25 metres) by its new president, a wealthy Parisian *avocat* named Ernest Archdeacon.

They knew better than to underestimate him, though, despite the revised appearance of his entry which, although it no longer included an airship and/or a donkey, was still extremely bizarre. (The configuration — tail ahead — later became known as a *canard* which, in French, means duck but also translates as hoax, false report and, for some reason, lump of sugar dipped in coffee.)

The judges and officials turned up twice at the Bois de

Ready to go and still looking extremely bizarre.

Boulogne, where he had once been rescued by the kite-flyers. The first time, after two runs, the ungainly combination of box-kites, now also sporting bicycle wheels, left the ground long enough for a huge cheer to go up before stalling and returning to the wet grass (on which the judges had been lying to get a better view) and shattering a propeller. Although it had travelled only a few metres, this was such an improvement on the usual nose-dives into the Seine that wild cheering continued.

Santos-Dumont was delighted. It was just a matter of a little more work on the fore and aft control, he told everyone, and then he would try again. Which he did, a month later, delayed

by bad weather and injuries to his arm in a balloon race.

At 8 a.m. on the morning of 22nd October 1906 everyone turned up again, including the press and photographers. Feeling as he did that "the calm of the dawn" was the most favourable time for flying, Santos-Dumont would have preferred to have seen them all earlier but he took it with good grace. "The duellist may call out his friends at that sacred hour," he declared, "but not the airship captain!"

After nine runs, a number of hiccups and a break for lunch, the moment came.

At 4 p.m. the 14-bis rose from the ground, described what was reported as "a graceful curve" and descended again to the

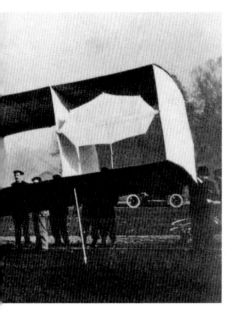

ground. The judges were so excited that they forgot to measure the exact distance but everyone agreed (and for this flight there were over a thousand witnesses) that it

Left: Although ailerons have been added, steering still looks like a bit of a problem.

Below: The wheels leave the ground.

"The first flight of a machine heavier than air. Mr Santos-Dumont winning the Archdeacon Prize."

THE ILLUSTRATED LONDON NEWS

was some 60 metres at a height of 2 to 3 metres. Although it wasn't the hoped-for 100, no one wanted to quibble. He had more than made the 25.

"Man has conquered the air!" flashed the headlines again. And again, the man was Santos-Dumont. The world went wild.

And three weeks later, he did it yet again.

Despite the celebrations and adulation, he had not been satisfied with making just 60 metres when 100 had been stipulated (there had been some muttering from his rivals on this

The crowd was a good size, but not half as large as retrospectively portrayed. Motor cars, prancing horses and even extra foliage were added to the scene.

962. M. SANTOS s'élevant avec son Aérsplane n° 14 bis (220 mètres en 21 secondes, 12 Novembre 1906) Surface 14 m. c, Moteur 50 HP. Poids total, aviateur compris, 300 Kgr J. H.

point) nor with the "graceful curve", which had actually resulted from a loss of lateral control.

On 12th November, two days after being hailed at an official commemorative banquet given by the Aero Club, he called the judges out once more.

To his surprise, they were not the only ones who turned up. Surrounded by a large crowd, another machine was on the field: the enormous biplane of Louis Blériot, which not only had a similar engine to the 14-bis but had also been constructed with the help of Gabriel Voisin. Should it make the 100 metres it could still win the combined prize and total glory.

Santos-Dumont always contended that prizes were of value more for the respect they conferred on aeronautics than for

4ᵉ. SPORTS - Aviation — " Santos-Dumont 14ᵇⁱˢ " (1906)

the actual money they offered,

which he didn't need – and often didn't keep.

With his legendary courtesy, he insisted that the intruder precede him, thereby forfeiting his chance to beat his own record. "Would Monsieur care to fire first?" he asked, and gallantly waited his turn until, after several runs, the challenging plane was totally wrecked without ever having taken off.

At this point, he hopped nattily into his wicker seat and, after a few preliminary lifts, rose into the sky, where, by means of samba-like movements which worked wires sewn into the back of his jacket – the world's first ailerons – he kept lateral control for 220 metres, and landed with no more damage than a gently collapsed wheel.

This time, consulting his namesake Cartier watch which

would become the first commercially sold wristwatch and, seventy-two years later, serve to remind the world of his feat, he knew he had done it. It had taken 21.2 seconds and set all doubts to rest, including his own.

The crowd rushed forth, lifted him once again from the gondola and carried him off, shoulder high, in victory, those who were near enough kissing and embracing him according to the French custom.

Non-stop celebrations ensued, he was awarded the 4,500 francs, and nothing was ever the same again. In addition to realizing his own personal dream, he had started mankind on a journey that would eventually lead it into space. He had created a more sympathetic climate for everyone from his rivals in Paris, muttering that the 14-bis was simply a freak, to the craziest oil-stained dreamer. And this included the Wrights.

In a very real way, he had opened the door for their claims which, although beginning to be talked about, had not been believed, even in their own country. They had, in fact, offered their designs for sale to the United States government, who had turned them down in 1905, as had the prestigious *Scientific American* in 1906.

Having left Kitty Hawk, they were now conducting experiments on a farm in Dayton, Ohio, observed mainly by a neighbouring farmer called Amos Stauffer, who used to watch them as he reaped his corn.

He didn't seem overly impressed.   *Never entirely satisfied.*

"Well, the boys are at it again," he frequently remarked to his hired man at the time, adding, "I just kept on shocking corn." And he wasn't the only one.

Their only champion seemed to be a reporter from a magazine called *Gleanings in Bee Culture*. What other publicity they received, in France and America, ridiculed them as dotty hicks.

HERE ON 12TH NOVEMBER 1906, UNDER THE CONTROL OF THE AERO CLUB OF FRANCE, SANTOS-DUMONT ESTABLISHED THE FIRST AVIATION RECORD IN THE WORLD. DURATION OF FLIGHT 21.2 SECONDS, DISTANCE 220 METRES.

An obelisk was set up at Bagatelle to commemorate the event.

Santos-Dumont, while he may well have heard of them by now, had no reason to think they were to change his life or that he had already helped to put that change in motion.

He spent a glorious winter, unrivalled and lionized, back at his favourite table at Maxim's, explaining his theories on aerodynamics to admirers and aspirants and, never entirely satisfied, working to improve the 14-bis. He was, as we might now say, on a roll.

CHAPTER ELEVEN

Dining with Louis Cartier

"I DID NOT YET KNOW MY EXACT TIME."

A ccording to the official Cartier archives, Louis-Joseph (who preferred to be known simply as Louis) Cartier met Alberto Santos-Dumont at an evening party organized by the Baron Deutsch de la Muerthe sometime in 1896.

Cartier, heir to the fortune and business founded by his grandfather in 1847, was, like Santos-Dumont, more than just another wealthy dilettante. In addition to their shared interests in motor racing, ballooning and evenings at Maxim's, both young men were perfectionists dedicated to fine craftsmanship and innovative design. Cartier, in fact, had been described – not altogether approvingly by traditionalists in his field – as "a dandy with revolutionary ideas".

The two respected, and supported, each other's

Louis Cartier

Cartier's rue de la Paix atelier, where early models of the Santos watch were sold.

work. Santos-Dumont ordered Cartier pieces for himself and his friends. Cartier turned up to witness the trials of Santos-Dumont's airships and balloons, applauding not only their ingenuity but also their craftsmanship (which he was not the only one to admire: the French internal Customs Office adjudged – and taxed – one of his designs as "fine cabinet work").

Himself somewhat reserved and enduring a difficult marriage which was later to end in divorce, Cartier also admired, according to the archives, "the free and easy manner of the rich Brazilian who conducted his life with total disregard for convention".

They saw each other frequently at motor races and other sporting events, and dined together often, at Maxim's and at each other's homes. Cartier would nonchalantly join Le Petit Santos at his elevated table where they were served from a little staircase by the valet, Charles.

Although Maison Cartier was already celebrated as the

leading personal jeweller to nobility, celebrity and power, Louis Cartier was, within his sphere, as interested in pioneering as his balloonist friend.

The advances of the Belle Époque, he felt, should certainly include those in the field of fine jewels, and he had been both stunned and spurred on by an exhibit in the Russian Building at the Grande Exposition. A huge map of France made completely of marble and studded with priceless stones, it perfectly expressed his feeling that the jeweller, was, as he saw himself, "a creator of works of art from precious materials".

He had been thinking for some time about the wristwatch which, since its first appearance in Tudor times (when Queen Elizabeth I's riding master, the Earl of Leicester, presented her with a "small watch with a bracelet"), had popped up every once in a while as what might be considered a ladies' novelty.

In the seventeenth century, nursing mothers took to tying their watches (normally worn around the neck on a chain) to their wrists with ribbon to keep their babies from grabbing at them. Gradually the idea grew, with women using ribbons, pearls and necklaces to fasten watches to their wrists for special occasions; the wealthier among them commissioned individual designs from their jewellers. Empress Josephine gave one to her daughter-in-law, Princess Amelia. Cartier himself had created a number, among them Santos-Dumont's gift to Belle de Neuilly. But no such inclination had been felt by men.

While watches for the wrist had been issued to the German

imperial army in the Boer War and, much earlier, the philosopher Pascal had astounded everyone by affixing one to his arm, in neither case had the idea taken hold. Men hung solidly on to their pocket watches.

When Cartier came up with the solution to his friend's bewilderment on landing after his great triumph and his complaints, later that evening at Maxim's, about the difficulty of keeping time aloft, he probably had no idea that he was creating what would become not only the first wristwatch for men, but also the best-selling (and most counterfeited) watch in the world.

"An elegant model," is how the archives recall it, "and the most famous one of all. Originally it had been conceived as a one-of-a-kind creation, a creation realized in the name of friendship, and far removed from all thoughts of war. The watch was called the Santos, and it was created by Louis Cartier during the Belle Époque, thereby sealing his friendship with Alberto Santos-Dumont, the Brazilian aviator and magnate."

While Santos-Dumont would have been surprised to hear himself so grandly referred to, he was the best person for whom the wristwatch could have been created. Not only did he inspire it, he also (more magnet than magnate) made it — like so much else he wore — fashionable.

Because it was a gift, it was never entered on the Cartier register, so the exact date of the making of the Santos is

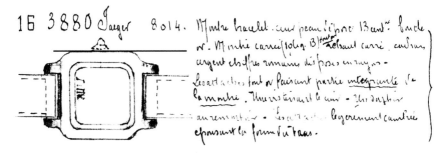

uncertain, but it was sometime after October 1901 and before November 1906 that the watch was delivered — personally (a great honour) — to the house on rue Washington. The generally accepted rough guess makes it 1904.

It was square (unlike all previous watches, which had been round), with Roman numerals, a second hand, a gold case, a sapphire winding crown and a simple, sturdy leather strap with a gold buckle. (What had been considered the original but turned out to have been a replica was later stolen from the Aeronautical Museum in São Paulo.)

Early models of the Santos watch (this one circa 1912) had a leather strap.

Cartier cuff links, and the gold bracelet holding the Comtesse d'Eu's St Benedict medal, were the only jewellery he ever wore.

The undoubtedly delighted Santos-Dumont wore it everywhere and, as he had with the Santos-Dumont collar, made everyone else want one, too. Early models were purchased privately – at the rue de la Paix atelier in 1911 and, in platinum, by Count Kinsky in 1913. By 1915, however, the Santos, described as "most popular", was commercially produced and remained a best-seller into the twenties and thirties when, like its inspiration, it gradually faded from sight.

Louis Cartier developed many other watches for famous clients, including an Easter egg for the Tsar and another, inserted in a hollowed-out, engraved, 136-carat cabochon emerald, for one of the American entertainers the Dolly Sisters. He made a giant tank watch for the Maharajah of Kapurthala and some watches small enough to fit into letter-openers, cuff links, cigarette lighters, lipstick cases and gold coins. He made watches with reversible faces, without hands and resembling sundials, and for everyone from King Farouk of Egypt to the Prince of Wales. But no other watch ever achieved the success of the Santos.

In 1978 it was reissued as the Santos Sport (*below*), with gold screws, stainless steel bracelet, and self-closing buckle, and enjoyed its second great success. Through it, the name of Santos-Dumont began to appear again – no longer only in professional journals (where the world of aviation had continued to revere him) but before the general public.

Advertising and promotional blurbs referred to "the renowned aviator pioneer" and his "daredevil exploits", and owners to this day themselves say they feel dashing wearing it. (See Author's Note.)

Made now in many models including gold, with diamonds, and for women, it has been judged to be not only Cartier's greatest success but an exemplary contemporary design and is on display in the permanent collection of the Musée de l'Air in Paris. Originally a product of the friendship between Cartier and Santos-Dumont, the watch has outlived them both.

CHAPTER TWELVE

Serpents in the Garden

"IT WAS, I MAY SAY NOW, RATHER PAINFUL FOR ME..."

Santos-Dumont flew the 14-bis once more, attaining up to 200 metres, then put it aside. It had proved its point (and went on to prove it once again when reconstructed in Brazil in 1956). By spring he had designed five new machines, including a plywood plane (another first), a hydroplane (which he floated but never flew), and one last airship-plane combination, before starting work on the plane that was to be his most loved creation, the airy little Demoiselle, or Dragonfly. Fashioned of bamboo and his favoured Japanese silk — which he insisted had to be yellow — she was both beautiful and fast, his crowning triumph, perfectly displaying the elegance of lightness and simplicity he had so long espoused. She was just 8 metres long, small enough to be transported in the back seat of his car.

As all over Europe men prepared to go one better than him, he spent a magical summer flying through the sunlight and

Santos-Dumont's last and most loved creation, the Demoiselle, the first light plane and one of the safest.

making delicate landings on the estates of friends just in time for lunch, which no longer had to be taken in a tree.

Crowds gathered just to watch the Demoiselle fly overhead and, once again happy to be the aerial sportsman, he continued the custom of throwing down his tie to them in a salute.

By autumn, though, things had begun to change. Some serpents had started to creep into the garden.

The heady, somehow innocent days of the Belle Époque were over. So, too, were those of early aviation when rivalry was accompanied by a kind of kinship.

Made of bamboo and
yellow silk from
Japan, the airy
Demoiselle was just
8 metres long.

*Santos-Dumont
enjoyed taking his little
plane out for a ride in
the back of his car.*

Now, everyone seemed to be taking to the air. Great air meets were beginning to be held in Rheims, as well as in Brescia, Italy, and in Manchester, England.

Competition was fierce.

Many of his colleagues were being killed in crashes.

Others were trying to take out patents on designs he continued to offer freely to all who wanted them. A car factory that had made some spare parts for the Demoiselle had to be legally prevented from trying to patent it. By the next year, happily – though with no profit to its creator except his

The Demoiselle, though delicate, was safe. No one was ever killed in a Demoiselle.

Living up to her name (meaning "young girl" or "dragonfly"), she was graceful and agile and could land on a lawn.

satisfaction — the Demoiselle was made in many countries including Germany and Argentina.

The first rumbles of the Great War were beginning to be heard.

And Wilbur Wright had come to Paris.

With an eye to selling his biplane to the French army, Wright began taking it up in Le Mans in August. Those who watched — and their numbers grew as word spread — were astonished at how readily it could be controlled.

While it was generally declared a superior machine to the 14-bis, no one, including Santos-Dumont himself, considered his reputation under threat. He never, as far as anyone knew, even went to see it.

The record had been set. He had been the first, and had remained unrivalled for over a year. That a number of other pilots (first Farman and, after several more crashes, later Blériot – both in planes built by Voisin) had already flown since was of no consequence. This was the natural progression of science.

In fact, shortly after the Wrights' Paris success, a better plane than theirs did appear. This impelled Santos-Dumont to point out that since its designer, Levavasseur, had been working on it for many years, he could have claimed that his had actually been the first. "What," asked Santos-Dumont, would have been said by Edison, Graham Bell or Marconi, if, after they had presented to the world the electric lamp, the telephone and the wireless, somebody else had turned up with a better lamp, a better telephone or a better

The Demoiselle was popular in many countries, including Germany and Argentina.

wireless and said that he had invented them long before?

Santos-Dumont (second from right) with friends at an early air meet.

"To whom then does the world owe flight in heavier-than-air machines? Is it to the Wright brothers who, according to their own account, kept the results of their experiment a carefully guarded secret, and were so little known to the world that my flight was described as 'a memorable moment in the history of aviation', or is it to Farman, to Blériot and to me who made all our demonstrations before scientific commissions and in the plain light of the day?"

And, while Wright's plane may have performed better than that of Santos-Dumont, Wright himself had most certainly not.

Stern in demeanour and spartan in his habits, he was

variously described in the press as "astonishingly phlegmatic" and lacking "*l'élégance et l'esprit*", the very qualities Parisians so loved in their Little Santos.

Wilbur Wright hated hotels, slept wrapped in a blanket beneath the wing of his plane, ate from tins, washed under a nearby hose, and – as befitted a former bicycle-repair man – travelled by bicycle. And perhaps it was these "macho" practices that prevented anyone from questioning the fact that, like Santos, he was seldom seen in the company of women.

He also had an irritating habit of whistling as he made his pre-test checks, spoke infrequently (although this may be put down to a lack of facility with the language) and, when he did say something, sounded not only stilted but dull.

"Gentlemen, now I am going to fly," was his usual pre-take-off remark and his response to the suggestion that this might be a bit too laconic was also oddly unsatisfactory. "The only birds that talk much are parrots and parrots don't fly very high," he declared, somewhat dourly and, as it happened, incorrectly. As Alberto Santos-Dumont could have told him, in Brazil, parrots soar high and brilliant through the sky.

There was a more important difference, however, between this tall, no-nonsense American and the small, voluble Brazilian: a self-proclaimed "businessman", jealously guarding his patents, he had come to sell. And his price was steep – $200,000.

He could, however, fly, and crowds began heading out to Le

The fickle press now pictured Santos-Dumont and Wilbur Wright in a duel.

Mans to watch him do it.

The change in attitude, though gradual, was steady.

Wilbur Wright became the new man of the hour and, as his celebrity burgeoned, so did the arguments. If he was the best, posited a growing number, why could he not also have been the first? A cartoon now appeared depicting Wright and Santos-Dumont in a jousting match.

Those against questioned why, if the Wrights had been first, they hadn't announced it at the time.

Those for cited the sophistication of Wright's plane as proof that they must have.

His old colleague Louis Blériot (left) never stopped believing in him.

Leaders in the field, particularly other pilots, continued to pay Santos-Dumont homage. "For us aviators," wrote Blériot, who had taken his own defeat with good grace, "your name is a banner. You are our pathfinder."

George Besançon, director of the Aero Club, was in no doubt. "What to us Parisians will always appear incredible," he declared, "when we remember the crowds that flocked to Bagatelle to see the first attempts of Santos-Dumont, is the fact that only five persons are supposed to have witnessed the famous success of the Wrights."

Public opinion, however, influenced by the excellence of their current plane, and led, as ever, by the press whose headlines had gone from "Fliers or Liars?" to "A decisive Victory for Aviation!" began to turn towards the Wright brothers.

While their hoped-for sale to the French government never went through, and it would still be some time before his record was officially disclaimed, Santos-Dumont found himself not so much toppled as eclipsed.

Ever the gentleman, he remained above the fray, although years later, after his death, a note was found in a locked drawer. "It was, I may say now," it read, "rather painful for me to note – after all my work in dirigibles and heavier-than-air machines – the ingratitude of those who a few years ago covered me in praise."

His supporters held fast. There are many to this day who still believe his record should stand. The Smithsonian long continued to refuse to exhibit the Wright brothers' plane for lack of proof (it spent twenty years in the British Science Museum) – and when they at last agreed, they were made to promise never to mount a rival claim. Finally, however, the Wrights' claim came to be the accepted one.

Aviation began to be portrayed as romantic.

While the question is complex, the main points of dissension as to the validity of the 1903 Kitty Hawk flight are that it was not properly witnessed, that it should have been reported and that it may not have been unassisted – a kind of launching pad was used which many confused with a catapult.

Whether, after the initial shock, Santos-Dumont might have fought back and somehow held his place in history will never be known, for he suddenly found himself under a more deadly attack.

CHAPTER THIRTEEN

One More Blow

"HOW MANY LIVES SACRIFICED...!"

H e had been feeling tired and nervous — symptoms that even then might have been attributed to the emotions of the moment — but when dizziness and double vision began to appear as well, he consulted a doctor. The eventual diagnosis was multiple sclerosis.

He was thirty-six years old.

The rest of his life was an anticlimax, spent wandering between France and Brazil, receiving honours and awards, and increasingly, as his symptoms worsened, in and out of sanatoriums.

Fastidious, proud and aware of the debilitation ahead, he closed up his apartment and airship station at once, pensioning off his faithful mechanic, Chapin, and instructing his valet, Charles, to give a message for his friends to the Brazilian Ambassador.

Santos-Dumont posing jauntily on the statue in his honour at St Cloud.

Place et Monument SANTOS-DUMONT. — Saint-Cloud.

For years, probably in order to avoid pity, he denied his illness, insisting that he had voluntarily retired. The disease itself, however, which is often accompanied by deep depression, was surely a factor in this and subsequent rash and isolating decisions.

He went first to the Normandy coast where he rented a small house, La Boîte, near Deauville and – poignantly, since he never flew again – took up astronomy. After a while he began receiving a few visitors, chief among them George Goursat and his nephew, Jorge, but he confided in no one, losing himself in meteorology by day and the stars at night.

He was by no means forgotten in Paris. He was now made a Grand Officer of the Légion d'honneur (its ultimate tribute), and several monuments (one depicting his hero, Icarus) were built to his achievements.

Even his old mechanic, Chapin, was given a special award, the Palme d'officier de l'Académie, normally reserved for writers, artists and scientists.

The papers, their interest renewed now that Wilbur Wright was no longer on the scene, ran frequent articles about his "temporary" retirement, predicting that he would soon return with some dazzling new invention. But he paid little attention to them, or to the growing intimations of the war to come.

It was with amazement, then, that he found himself accused of being a spy by the local residents, who knew only that he was a foreigner with a German telescope on his roof,

constantly looking out at the sea.
Assuming that he – he who had freely
offered his entire aerial fleet to the
government of France – was sig-
nalling to U-boats, they had his house
searched by the police.

Although the mistake was soon discovered and official apologies
offered, his reaction was extreme.
Bitter and heartbroken, he burned all
his papers – destroying the very notes,
letters, designs, and even a diary, that
could have helped to keep his memory
alive – and returned to Brazil where
he spent the war years, mostly in despair.

The Queen of Romania meets the heroic Sportsman of the Air.

Increasingly unstable, with his old conflict never resolved,
he blamed himself for the dropping of bombs from zeppelins,
airships and planes.

From time to time, following the typical pattern of the illness which was, of course, even less understood then than it is
now, there were periods during which its effects abated, allow-
ing him to sail his little boat, the *Tico-Tico* (Sparrow) and build
himself a house and another observatory on a hillside above
Rio. He called it La Encantada, the Enchanted Place, and filled
it with gadgets, including a half-step staircase, which he had
invented to aid his steadily decreasing mobility. He wrote

another book, *O Que eu vi, O Que Nos Veremos* (Where We Have Been, Where We Are Going), a depressed, confused account, with none of the joyful exuberance of *Dans l'air*.

Still held in high esteem, he was invited to Washington, where he addressed Congress, albeit somewhat ramblingly, was made first president of the newly-formed Pan American Aeronautical Foundation and served as the delegate of the United States to its conference in Chile.

These were among his last good moments. Returning to France after the war, much aged and steadily deteriorating, he found no peace, even in Paris. He wandered from country to country including England where, in 1922, he spent some time staying at the house of a friend in Bath. It was here, hanging from the ceiling (suspended on wire, of course), that a model of a forerunner of

His illness put a halt to work. He closed up his apartment, and never flew again.

He designed an "enchanted" house, La Encantada, and devised gadgets to aid his decreasing mobility.

the 14-bis was noticed by Mr H. Jephcott-Tanburn, OBE.

Explaining that it hadn't been taken down in more than twenty years, Santos-Dumont, retaining his desire to amaze despite his illness, then not only took it down but made a present of it.

Mr Jephcott-Tanburn, in turn, presented it to the RAF Club where it stayed for another twenty-five years (during which it was damaged while being played with by some unnamed members) before being moved to the Science Museum. Thereafter it was purchased by Vickers-Armstrong Aircraft and, in 1957, presented to the Air Ministry of Brazil in commemoration of the fiftieth anniversary of the flight.

Despite such mini-replays of earlier times, Santos-Dumont

finally became too restless to stay in any one place and, with no family to contain him, he entered a sanatorium in Switzerland.

He had been there a year when the Aero Club invited him to preside over their gala banquet for the American aviator Charles Lindbergh, who had just flown across the Atlantic. As he wrote, weeping, that he was too ill to accept, his hand shook so that his signature, once so proud and flowery, was only a feeble scrawl.

His decline continued, made all the sadder for the failure of his few remaining attempts to carry on. His petition to the League of Nations to ban further aerial warfare was ignored and his penultimate invention, motor-driven wings made of swan's feathers and wire, was based on an idea he had rejected as foolish in his early years. He did, however, rally enough to scrap it and to use the engine to power something more in the "Little Santos" spirit. The result – a kind of individual device to help weary skiers up the slopes (the lift had yet to be invented) – had

Perhaps the last Santos-Dumont creation, a motor-driven precursor of the ski-lift..

After the accident, a distraught Santos-Dumont sought solace at the beautiful Copacabana Palace Hotel.

some slight success at St Moritz.

Papers and magazines had to be kept from him as he began to feel personally responsible for every plane that crashed.

Finally, he was so ill that Jorge came to take him back to Brazil — where fate had one more blow in store.

Having already awarded him every other honour they could think of, including his name on banknotes and coins and the restoration of his birthplace, the government planned to elect him to the Brazilian Academy of Letters.

A reception committee, half literary, half aeronautical, awaited his ship as it sailed into the bay of Guanabara on 3rd December 1928.

A hydroplane, rechristened the *Santos-Dumont* for the event and carrying the cream of the country's intelligentsia, flew out to drop a parachutist with a welcoming proclamation. Instead, as he stood watching on the ship's deck, the hydroplane blew up and crashed into the sea, killing everyone aboard. For him, it was the worst possible thing that could have happened, the ultimate embodiment of his recurring horror.

"How many lives sacrificed for my humble self," he kept repeating.

He insisted on trying to help retrieve the bodies, attended each funeral and then shut himself up in the newly-opened Copacabana Palace Hotel (where his signature, dated 29th December 1928, is recorded in the Golden Book, just above that of John Galsworthy), before retreating to his enchanted house above the city.

His signature can still be seen in the Copacabana Palace Golden Book.

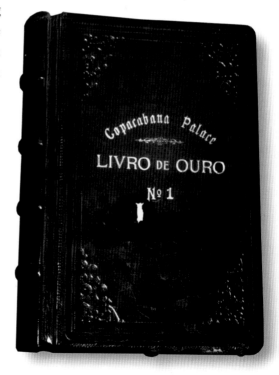

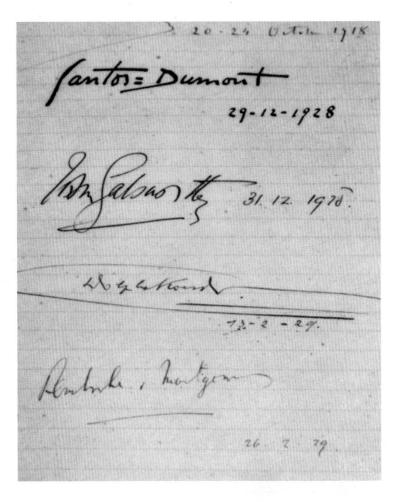

In addition to his signature, just above John Galsworthy's, his photograph remains on display — alongside those of John Wayne, Nelson Mandela, Brigitte Bardot and Prince Charles.

The spell had been broken, however, and he began wandering again. Finally, as his lucid moments grew fewer and he lost more and more control of his functions, he settled by the big, broad beach at Guarujá in the care of the dutiful Jorge, whom he had once visited in Tavistock.

Here he had some peaceful days, watching the children play on the shore, and even, in his better moments, walking along it himself.

He took pride in the peacetime use of the big dirigibles and saw a dream come true when the *Graf Zeppelin*, featuring roast gosling and sleeping cabins, flew from Germany to Brazil, its wealthy and glamorous passengers sprayed, "the ladies lightly with eau de Cologne, and the gentlemen with soda water", in a scaled-down Aeolus ceremony. But what he saw as the disastrous results of his life's work followed him even there.

Civil war broke out in Brazil; friends were imprisoned or banished, his own countrymen killing one another with the weapons he had created. Once again he made a plea for peace; once again it went unheard.

CHAPTER FOURTEEN

The Morning of 23rd July

O n the morning of 23rd July 1932, three days after his fifty-ninth birthday and two weeks after the fighting had begun, he watched as planes flew low over the broad, beautiful beach, then heard the sound of dropping bombs.

Overcome by melancholy and disease, he despaired that his inventions had been twisted to such a ghastly end. His reason snapped at last.

Choosing a tie of the sort he had use to throw to the cheering crowds, he went into the bathroom and, with his palsied hands, looped it around his neck and over a hook, and hanged himself.

There was a state funeral (his death ascribed to cardiac causes) and the world

His memorial at Rio Airport.

mourned – even the Civil War ceased for two days (although its leaders had actually promised to stop the fighting for three). He was buried in Rio, beneath a statue of Icarus. His estate was small. He had spent most of it on his inventions and, taking delight in offering them freely to mankind, had received little back.

It was, in fact, partly his very generosity that led to his eclipse. Although the possibility certainly couldn't have occurred to anyone at the time as the tributes rolled in, his memory, lacking patents, backers, children or even documents to keep it alive, simply faded away.

"He left us, as a legacy, nothing but his name, engraved on our hearts," wrote Gabriel Voisin – but he was wrong on both counts.

"Those who knew him could not help but love him."
Gabriel Voisin

CHAPTER FIFTEEN

Man Flies

"LET GO ALL!"

The name was not a lasting legacy. It was gradually forgotten, engraved on few hearts but those of balloon aficionados and in Brazil, where he is still a national hero. Until 1978, when Cartier revived the watch and, with it, some shred of his reputation, his name could be seen only in a few dusty archives and on some statues and street signs around the world.

But it wasn't his only legacy. He also left us his accomplishments, which remain with us to this day.

He was the first to make a powered, controlled flight of any kind, showing that man could navigate the air. In that sense alone, he truly was, as he was so often called, the Conqueror of the Air, the Father of Aviation.

He also built the world's first light plane, and one of the safest — no one was ever killed in a Demoiselle.

As to other records, he certainly made the world's first

The Desmoiselle survives on a current postcard.

public, timed and judged heavier-than-air flight, the first such European flight and, in the Demoiselle, the first cross-country flight.

Designing and building all the machines for these feats himself, he performed experiments, including those with wheels, gyroscopes and ailerons, which, patented or not, directly affected all future aeronautic development.

Although emphasis on his style has sometimes obscured the value of his contributions, he was — and is — certainly as important as Lindbergh, the astronauts and his own early heroes, and more important than many whose names have lasted longer than his.

Whether he technically made the first ever heavier-than-air flight is, in a way, irrelevant because for several years the entire world believed that he had, and was inspired by his spirit which, in the end — and at the end of the millennium — may be the thing that will never be matched.

Ignoring ridicule and making it look achievable, he encouraged others to aim for the stars.

Wherever man flies, the spirit of Alberto Santos-Dumont, "Little Santos", flies with him.

The Airships

BRAZIL   1898

Spherical free balloon; volume 13 cubic metres

No.1   1898

Elongated, non-rigid airship; petrol engine (made from motor-cycle parts) and propeller mounted on passenger basket; directional control by rudder, longitudinal by moving ballast weights; length 25 metres

Air pump was insufficient, causing balloon to double up

No.2   1898

Non-rigid airship; similar engine to No.1; same controls

Cold rain caused the hydrogen to contract; balloon crashed into trees

No.3   1899

Semi-rigid; similar engine and many fittings; same controls; length 20 metres

Successful – but lost rudder on last flight

No.4   1900

Semi-rigid; new engine; same controls;
water ballast; engine, tanks and pilot (on
bicycle seat) on longitudinal pole; length
29 metres

*Not very successful – spider's web configuration of
ropes compromised the shape*

No.5   1901

Semi-rigid; Buchet air-cooled engine based
on Daimler-Benz design; braced with piano
wire; girder keel; length 34 metres

Crashed in Rothschild garden and into Trocadero Hotel

No.6   1901

Semi-rigid; Buchet/Santos-Dumont water-
cooled engine; same fittings as No. 5;
length 33 metres

*Won the Deutsch Prize for a timed controlled flight
around the Eiffel Tower; later crashed in Monaco and
had to be rebuilt*

No.7 the Racer   1902-4

Semi-rigid; Clément engine; propellers
front and rear; length 40 metres

Never raced

No.8   NO NUMBER 8

No.9 La Baladeuse (the Stroller)   1903

Semi-rigid; Clément engine; length
11 metres
*Used for leisurely jaunts; backfired (*retour de
flamme*) over the Seine (panama hat incident)*

No.10 the Omnibus   1904

Semi-rigid; engine and controls interchange-
able with No.7; four little baskets designed
to carry total of up to ten passengers; length
42 metres
Tests only

No.11   1905

Glider, towed behind boat; redesigned as
twin-propeller aircraft
Never given power

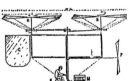

No.12   1905-6

Helicopter mock-up
No suitable engine, so never flown

No.13   1905

Semi-rigid airship; carried own hot-air
generating plant
Never flown

No.14   1905

Semi-rigid; Clément engine; test vehicle
used to lift heavier-than-air craft

Abandoned as "monstrous hybrid" and used simply as
airship

No.14-bis   1905-6

Biplane *canard*; Levavasseur/Antoinette
engine; controlled by forward box-kite;
ailerons fitted in later modification; area 42
square metres

Won the Archdeacon Prize for the first heavier-than-
air flight

No.15   1906-7

Plywood biplane; Antoinette engine

Crashed on first test; never airborne; never rebuilt

No.16   1907

Semi-rigid airship with wings; Antoinette
engine; meant to combine lighter- and
heavier-than-air

Destroyed on ground before flight

No. 17   1907

Biplane

Designed but never built

No. 18   1907

Wingless hydroplane; Antoinette engine

Floated but never flown

No. 19   1907

Monoplane; Dutheil / Chalmers engine
modified by Santos-Dumont; bamboo and
silk; prototype for Demoiselle

*Never fitted with wings; damaged in early tests on
water*

No. 20 Demoiselle (Dragonfly)   1907

Modification of No. 19; length 8 metres

A delight; Santos-Dumont's favourite; first light plane

No. 21 Demoiselle   1909

Antoinette or Darracq engine

No. 22 Demoiselle   1909

Bayard engine

Chronology

1873   Alberto Santos-Dumont born in Brazil (20th July)
            death of Napoleon III
            lawn tennis introduced in England
            abolition of slave markets in Zanzibar

1891   Santos-Dumont moves to Paris
            Gaugin goes to Tahiti
            adventures of Sherlock Holmes appear in *The Strand* magazine
            invention of the zipper

1896   Santos-Dumont meets Louis Cartier
            Nobel Prize established
            Klondike gold rush begins in Canada
            Pithecanthropus Man discovered in Java

1898   Santos-Dumont takes petroleum motor up in tree; creates the
                   Brazil and No. 1
            birth of Ernest Hemingway and Bertolt Brecht
            Paris métro opens

1899   Elgar wins recognition with *Enigma Variations*

1900   Grande Exposition begins

Freud's *The Interpretation of Dreams* published
birth of HM Elizabeth the Queen Mother

1901   Santos-Dumont wins Deutsch Prize in No.6
              (19th/23rd October)
        Rasputin gains influence in court of Tsar Nicholas II
        adrenalin discovered
        ragtime at its height

1903   Santos-Dumont builds No. 9 (La Baladeuse)
        Henry Ford founds Ford Motor Company
        electrocardiograph developed
        Richard Steiff makes first teddy bear

1904   first Santos watch created (?)
        birth of Marlene Dietrich
        Charles Rolls meets Henry Royce

1905   Einstein's special theory of relativity formulated

1906   Santos-Dumont flies 14-bis (23rd October); wins
              Archdeacon Prize (12th November)
        San Francisco hit by great earthquake
        birth of Mary Lois Miller

1907   Santos-Dumont builds Demoiselle
        Pavlov experiments with conditioned reflex
        first motor taxis in London
        Baden-Powell founds Boy Scout movement

1908   Wilbur Wright's first French flight (Le Mans)
            The Wind in the Willows published

1910   Santos-Dumont taken ill
            tango sweeps Europe and United States

1911   Santos watch sold commercially
            Rupert Brooke's *Poems* published
            death of W. S. Gilbert
            temperature in London reaches 100° Fahrenheit

1922   Santos-Dumont living in England near Bath
            Gandhi jailed for conspiracy
            F. Scott Fitzgerald writes *The Beautiful and Damned*

1928   hydroplane *Santos-Dumont* crashes into Brazilian bay
                (3rd December)
            Walt Disney creates Mickey Mouse

1932   Santos-Dumont hangs himself with tie (23rd July)
            Hitler defeated by Hindenburg in presidential elections
            Lambeth Bridge completed; Golden Gate Bridge begun
            Alexander Calder exhibits mobiles
            Zuider Zee reclaimed

1978   Santos watch reissued as Santos Sport

23rd October is celebrated every year throughout South America as
Aviators' Day

Glossary

AERONAUT originally a practitioner of any branch of aviation, but principally applied to balloonists

AEROSTAT a balloon; aerostatic lift is through buoyant gases, as opposed to dynamic lift by movement through the air

AILERON movable control flap on aircraft's wing

AIRSHIP a lighter-than-air craft that has propulsion and steering; synonymous with steerable balloon/dirigible; three main types: rigid, non-rigid and semi-rigid

"AWAY ALL ROPES" a call for all ground crew to release their hold ("Hands off" is more generally used in ballooning)

BALLAST usually sand or water, although sometimes lead shot, released to make a balloon or airship lighter

BALLOON a craft that flies by means of aerostatic lift (the part that contains the gas is more correctly known as the envelope)

BIPLANE an aeroplane with two pairs of wings, one set above the other

DIRIGIBLE an early word for a steerable balloon or airship

GONDOLA the boat-like cabin beneath an airship

GUIDE-ROPES handling lines for manœuvring airships or balloons from the ground

GYROSCOPE a device with an internal rotating flywheel which establishes directional and vertical reference

HEAVIER-THAN-AIR applied to aircraft that obtain lift through dynamic means, pushing through the air

HOT-AIR the original form of balloon flight; rediscovered in the 1960s by the marriage of new fabrics with powerful propane gas burners

HYDROGEN lighter-than-air gas; easy to manufacture, but extremely explosive if mixed with air (helium is the modern safe alternative, but very expensive)

HYDROPLANE aeroplane with floats or a sea-going hull for water landings; nowadays more often known as a seaplane or floatplane

INTERNAL COMBUSTION ENGINE a petrol or diesel engine; the first power source that was light enough to be useful to airship builders

KEEL as on a boat, the strengthening structure running along the bottom of

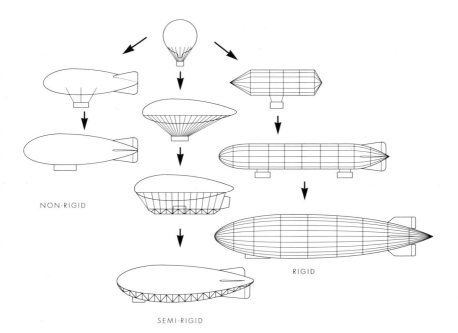

NON-RIGID

RIGID

SEMI-RIGID

a rigid or semi-rigid airship; sometimes suspended slightly below the hull

LIGHTER-THAN-AIR applied to aircraft that fly through aerostatic lift, by gas

MONOPLANE aeroplane with a single pair of wings

NON-RIGID AIRSHIP an airship with no internal framework, often called a

blimp or pressure airship, which maintains shape through internal gas pressure alone

RIGID AIRSHIP an airship with an outer framework of aluminium or even wooden girders surrounding a balloon or cell of gas

SEMI-RIGID AIRSHIP an airship that combines a pressure airship with a rigid keel

to distribute the load

STEERABLE BALLOON an early term for an airship

VALVING releasing gas to make a balloon descend

ZEPPELIN a rigid airship; mainly applied to those built by the German Zeppelin company

Bibliography

Aimé, Emmanuel, "La Navigation Aérienne au XXe Siècle", in *Revue Ampère* (November 1901).

Anon, "Santos-Dumont: The Conqueror of the Air", in *Brazil Today* (1954).

Botting, Douglas, *The Giant Airships* (Time-Life Books, Virginia, 1980).

Dollfus, Charles, and Bouché, Henri, *Histoire de l'Aéronautique* (Paris, 1942).

Gautier, Gilbert, *Cartier: The Legend* (Arlington Books, London).

Gibbs-Smith, C.H., *A History of Flying* (Batsford, London, 1953).

Napoleao, Aluizio, *Santos-Dumont and the Conquest of the Air* (National Printing Office, Rio de Janeiro, 1945).

Santos-Dumont, A., "The Sensations and Emotions of Aerial Navigation", in *Pall Mall Magazine* (1904).

Santos-Dumont, A., *Dans l'air*, translated by Peter Smith as *My Airships* (Dover Publications, New York, 1973).

Souza, Marcio, *O Brasileiro Voador* (Rio de Janeiro, 1986).

Villares, Henrique, *Santos-Dumont: The Father of Aviation* (São Paulo, 1956).

Vreeland, Diana, *D.V.* (da Capo, New York, 1997).

Wohl, Robert, *A Passion for Wings* (Yale University Press, New Haven, 1994).

Wykeham, Peter, *Santos-Dumont: A Study in Obsession* (Putnam, New York, 1962).

Acknowledgements

Many thanks for their various roles in getting *Man Flies* off the ground to: Bill San Filippo, Buddy Bombard, Tucano, Roberto Maksoud, Maryann Bowen, Mark Barty-King, Mike Shaw, Teresa Buxton, Denis Billecourt, the man on stilts, Penny Phillips, Alan Noble, Alan Parker, Bonham's, John Christopher, Robin Batchelor, Peter Mossman, Ali Gunn, Thom Carver, Jonathan Thornton, Venita Paul, Dr Shirley Sherwood, Claudia Failho, Lionel Lambourne, Rotisserie Jules, Roy Williams, Graca Fish, Keith Patrick, Terry Davidson, Léon Siegler and the Away All Ropes team, especially Polly Napper, Jocasta Brownlee, Helena Drakakis, Noni Ware, Suzie Yuan, Steve Blackburn and Catherine White. Thanks also to the Science Museum, Royal Aeronautical Society, Aeronautical Museum Iribuera Park, Air Ministry, RAF Museum and Cartier.